TILT HAPPENS

The Definitive Guide to Dysautonomia Testing

By: **Dr. Khalid Saeed, D.O.**

Legal Stuff (Because Apparently Gravity Isn't the Only Thing With Rules)

Medical Disclaimer
This book is intended for educational, informational, and aggressively sarcastic purposes only. It is not medical advice. It is not a substitute for personalized evaluation, diagnosis, or treatment. And it will absolutely not convince your clinician to suddenly understand autonomic physiology if they have spent the last decade assuming the sympathetic chain is a fictional creature that lives behind the spleen and grants wishes.

Even though the author, Dr. Khalid Saeed, D.O., is a licensed physician who also has dysautonomia (lucky him), he is not your doctor unless you've met him in person, filled out enough intake forms to qualify as a graduate thesis, and discussed your cardiovascular quirks without needing to adjust the exam room temperature.

Always consult your own licensed healthcare provider before making any changes to your medications, salt intake, fluid strategy, exercise plan, neuromodulation routine, cooling device collection, or life choices—especially if you are pregnant, nursing, taking medications, or currently horizontal and negotiating with gravity.

Reading this book does not create a doctor–patient relationship. Snark is not a billable service. And while autonomic dysfunction is real, measurable, and occasionally dramatic, this book exists to give you clarity—not to replace clinical care. The author makes no representations or warranties, express or implied, regarding the completeness or accuracy of the content.

Copyright Notice
© 2025 Khalid Saeed, D.O. / *Tilt Happens: The Definitive Guide to Dysautonomia Testing*™ / ISBN 979-8-218-88422-2

No part of this book may be copied, reproduced, reprinted, scanned, screenshotted, uploaded, memed, rearranged into interpretive dance skits, fed into an AI model, or used to shame your doctor during an appointment without prior written permission from the author—except for brief excerpts used in reviews, commentary, education, or strongly worded social media posts that properly cite the source.

This work is protected under the Copyright Act of 1976 (Title 17, United States Code) and international copyright treaties. Unauthorized reproduction may result in legal action, autonomic misfortune, and the universe assigning you a lifetime supply of compression socks in a color best described as "clinical despair."

Patent Notice
Portions of the diagnostic frameworks, analytical methods, interpretive models, and system architectures described in this book are the subject of one or more U.S. patent applications. This includes content covered under U.S. Provisional Patent Application No. 63/935,938, filed on 12/10/2025. A corresponding non-provisional application has been or will be filed claiming priority to that provisional application. Certain methods and systems described herein are patent pending.

Trademarks, Satire, and Nominative Fair Use
This book contains satire, parody, commentary, and criticism about (but not limited to) wellness trends that forgot about physiology, hydration culture, the "just stand up slowly" industrial complex, wearable devices with the confidence of PhDs, and clinical behaviors that suggest some people skipped the autonomic chapter entirely.

All brand names, product references, medical devices, diagnostic terms, institutional names, and public health slogans remain the property of their respective owners. Their inclusion is purely nominative, educational, or satirical. Nothing here implies endorsement, sponsorship, collaboration, or that the author and the trademark holders would get along at a conference. All trademarks are used solely to identify the products or services to which they refer.

Use of trademarks falls under nominative fair use as recognized by New Kids on the Block v. News America Publishing, Inc., 971 F.2d 302 (9th Cir. 1992) and the Lanham Act, 15 U.S.C. § 1115(b)(4). Portions of this book also fall under fair use as outlined in 17 U.S.C. § 107 and reinforced by Campbell v. Acuff-Rose Music, 510 U.S. 569 (1994). Nothing in this notice is intended to restrict lawful fair use rights under 17 U.S.C. §107.

In normal-person English, it's legal to mention brands or concepts when the point is to educate, critique, or clarify—and no sane alternative exists that wouldn't confuse readers already dealing with an autonomic plot twist every time they stand up.

Liability Disclaimer
The author and publisher disclaim all liability, loss, injury, or damage—physical, emotional, or existential—that may arise from reading this book, attempting to self-diagnose, lecturing your doctor using newly learned baroreflex terminology, or trying to put on pants while your baroreceptors send out ransom notes.

Your health decisions are your own responsibility. Use caution. Use logic. Use your clinician. If you're using this book as a replacement for an actual treatment plan, please lie down immediately and reevaluate your choices with professional help. Snark is not a scope of practice. Physiology is.

ACKNOWLEDGMENTS

(The part where I pretend I wrote this book alone.)

First, to everyone living with dysautonomia—thank you for turning survival into a science. Your collective persistence, ingenuity, and ability to navigate daily life with a nervous system that behaves like a moody software update made this book possible. You deserve better explanations, better care, and fewer people suggesting "just hydrate." This book is for you.

To every patient who tolerated being told "your tests are normal" while their physiology was doing a flash-mob dress rehearsal backstage—thank you for inspiring the frameworks, the snark, and the refusal to accept vague answers. Your experiences shaped every chapter.

To the clinicians who *actually* measure physiology instead of guessing—you are the rare unicorns of modern medicine. May your tilt tables always function, your Phase IV returns always make sense, and your colleagues stop confusing "upright intolerance" with "attitude."

To the clinicians who dismissed dysautonomia as "anxiety"—congratulations. You helped fuel an entire book out of sheer irritation. Your contribution is noted, archived, and deeply appreciated in a way you may not enjoy.

To the autonomic researchers who publish brilliant work that only ten people read—I read it. I understood at least seven percent of it. Thank you for giving the field real mechanisms, even if the world hasn't caught up yet.

To every friend, colleague, and unsuspecting bystander who listened to me explain preload, baroreflex timing, and the Valsalva maneuver at socially inappropriate moments—your patience was heroic. Your blank stares were motivating.

To the chairs, sofas, benches, stools, and other horizontal surfaces that hosted me while writing this book—your service to science has not gone unnoticed.

To gravity—you remain undefeated, but not unchallenged.

And finally, to everyone who trusted me enough to read this book—clinicians, patients, skeptics, and those simply trying to understand why standing up feels like a contact sport—thank you. You're the reason I wrote this, the reason I keep going, and the reason the autonomic world is evolving, one snark-filled explanation at a time.

Tilt happens. But so does progress. And you're part of it.

DEDICATION

For the people who live in a world designed for someone else's physiology.

To everyone with dysautonomia who has ever sat down on the floor of a grocery store, leaned against a wall pretending it was intentional, carried a personal microclimate's worth of cooling gear, timed a meal like a biochemical event, or explained—again—that no, this isn't anxiety, it's physics. You're the reason this book exists.

To the patients who kept searching for answers long after they were told there were none: your persistence built this field more than any institution ever has.

To the clinicians who actually measure physiology instead of guessing: you give me hope for the future of medicine. Also, you deserve more funding.

To the people who taught me—directly or accidentally—how confusing, frustrating, and absurd dysautonomia can be: thank you for shaping the voice, the humor, and the fire behind every page.

And to my own autonomic nervous system, which insisted on becoming a full-time collaborator: you could have sent an email. But here we are.

This book is for every person who has been tilted—by gravity, by symptoms, by dismissal, by doubt—and kept moving anyway.

Tilt happens. But so does defiance. And this is dedicated to you.

TABLE OF CONTENTS

Acknowledgments i
Dedication iii

INTRODUCTION: The Autonomic System, Cornered at Last 9
PART I: KNOW YOUR ADVERSARY: THE AUTONOMIC
SYSTEM UNMASKED 15
    CHAPTER 1: The Autonomic Nervous System: The Original
    Multitasker 17
    CHAPTER 2: Heart Rate Variability: Listening to the Heart–
    Brain Conversation (Even When They're Arguing) 27
    CHAPTER 3: Breathing, Bearing Down, and Standing Up:
    The Core Autonomic Reflexes 37
    CHAPTER 4: When the Skin Talks: Sudomotor and Small Fiber
    Function 47

PART II: THE WEAPONS ROOM: HOW WE INTERROGATE
YOUR NERVOUS SYSTEM 57
    CHAPTER 5: Resting State: The Autonomic Baseline 59
    CHAPTER 6: The Deep Breathing Test: The Parasympathetic
    Playbook 69
    CHAPTER 7: Valsalva Maneuver: A Sympathetic Stress Test
    Without a Treadmill 79
    CHAPTER 8: Stand or Tilt: Orthostatic Warfare 89
    CHAPTER 9: Sudomotor Electrochemistry: Sweat as a Diagnostic
    Tool 101

PART III: THE INTERPRETATION ENGINE: HOW TO READ THE BATTLEFIELD 111

CHAPTER 10: Normal-Normal, Normal-Abnormal, Abnormal-Abnormal 113

CHAPTER 11: Medication Interference: When Pharmacology Wears a Fake Mustache 123

CHAPTER 12: Heat, Food, Stress, and Standing in the Real World: Functional Triggers 133

PART IV: MECHANISMS EXPOSED: IDENTIFY YOUR ENEMY 143

CHAPTER 13: Test Integration: Identifying the Dominant Failure Mode 145

CHAPTER 14: Prognostic Indicators: What Predicts Progression vs. Recovery 159

PART V: THE COUNTEROFFENSIVE: MANAGEMENT BY PHENOTYPE & MECHANISM.. 171

CHAPTER 15: POTS: Precision Treatment by Subtype 173

CHAPTER 16: Mechanism-Directed Therapy 187

CHAPTER 17: Advanced Management Strategies 201

CHAPTER 18: Clinical Case Integration 211

PART VI: REAL-WORLD SURVIVAL: WINNING IN HOSTILE CONDITIONS 223

CHAPTER 19: Environmental Warfare: Surviving Heat, Food, Travel & Gravity's Friends 225

CHAPTER 20: Rehab Without Collapse: Rebuilding Autonomic Stability Without Triggering Physiologic Mutiny........................ 233

CHAPTER 21: The Autonomic Test Report as a Weapon: How to Interpret, Integrate, and Actually Use the Data 241

CHAPTER 22: Action Maps & Clinical Rapid Reference:
Turning Mechanisms Into Medical Decisions........................249
CHAPTER 23: Real-World Adaptation & Safety Protocols:
Clinician-Guided Strategies for Preventing Physiologic Ambush....259
CHAPTER 24: Future Directions of Autonomic Medicine..............267

CONCLUSION: The Gravity-Defying Truth............................275
**FINAL THOUGHT: The Last Word From Both Sides of
the Exam Table** 279

**ABOUT THE AUTHOR: The Clinician Who Read the
Fine Print**..281
CREDITS..283
REFERENCES...285
SUGGESTED READINGS & TOOLS 291
GLOSSARY OF TERMS..297
APPENDIX: THE RAPID-REFERENCE ARSENAL...................309

INTRODUCTION:
THE AUTONOMIC SYSTEM, CORNERED AT LAST

(Where we stop pretending dysautonomia is a mystery and start admitting it's a full-scale physiological mutiny.)

Welcome to the only part of medicine where gravity wins more often than it should, where blood vessels forget their job descriptions, and where the nervous system decides—unexpectedly—that today is simply not the day it will cooperate.

If you're reading this, you've either lived through that circus, treated someone mid-performance, or are finally ready to figure out why a fully grown human being can be taken down by standing upright, heat, food, stress, or the unspeakable challenge of taking a shower.

This book is here to tell you a secret the body has been screaming for years—the autonomic nervous system (ANS) is running everything, and it is deeply offended that no one has bothered to learn its rules.

The ANS controls heart rate, blood pressure, thermoregulation, digestion, immune modulation, endocrine balance, perfusion, energy allocation, and basically every function the body considers too important to entrust to your conscious mind.

It is the backstage crew keeping your physiology from collapsing in public, and like any backstage crew, it only gets attention when something goes catastrophically wrong—when the lights fall, the wires smoke, and suddenly everyone wants to know who's in charge back there.

Well, this book answers that. And it answers it with precision, science, sarcasm, and absolutely no patience for diagnostic folklore.

Why This Book Exists

Autonomic dysfunction has been mislabeled, misinterpreted, misdiagnosed, and dismissed with a level of clinical confidence only achievable by someone who has never actually looked at autonomic physiology.

Patients are often told: *your vitals are normal, it's probably stress, drink more water, it's anxiety, you just need to exercise,* and *try mindfulness.*

Meanwhile, their autonomic nervous system is backstage waving a flaming baroreflex and yelling, "CAN ANYONE SEE THIS?"

This book exists to flip the script.

Not to sympathize—to clarify.
Not to theorize—to measure.
Not to guess—to test.

And yes, to roast physiology when it misbehaves.

What This Book Is Not

This is not a book about vibes.
This is not a book about intuition pretending to be data.
This is not a book about being told everything is fine.
This is not a book that tells patients their symptoms are feelings in disguise.

This is a book about testing, mechanisms, compensation curves, failure modes, and physiologic triage—and the uncomfortable truth that the autonomic nervous system is only intuitive to people who truly understand it, which excludes far more clinicians than anyone is comfortable admitting.

You're going to see the ANS exposed, interrogated, plotted, analyzed, and occasionally mocked.

What This Book *Is*

A full breakdown of autonomic testing—what it means, how it works, and why it diagnoses what surface-level vital signs never will.

You will learn:

- What "normal" actually looks like and why many patients have never experienced it.
- Why HRV is not a wellness trend but an early warning radar with no off switch.
- How deep breathing, Valsalva, and standing up reveal more than most lab panels combined.
- Why sudomotor testing is the autonomic version of a lie detector.
- How to interpret patterns that actually differentiate POTS subtypes, orthostatic hypotension, baroreflex failure, neuropathic states, and central instability.
- How medications disguise or distort physiology—sometimes spectacularly.
- How to extract mechanisms from chaos and then use them to create targeted treatment plans.
- How to navigate real-world triggers like heat, meals, and stress without collapsing in front of your refrigerator.
- How to wield your test report as a weapon instead of a curiosity.

This book is not merely informative. It is weaponized clinical literacy.

Who This Book Is For
Patients: You've been dismissed, misdiagnosed, misinterpreted, and under-tested.

This book hands you the toolkit you should have had from the beginning.

Clinicians: If you've ever looked at a dizzy patient and confidently declared, "Vitals are normal," this book is your redemption arc.

Researchers: You already know the ANS is chaos on parade. This book organizes it.

Anyone who has fainted, nearly fainted, or thought about fainting: Welcome to the club. Gravity doesn't discriminate.

How to Read This Book (Without Experimenting With Gravity)
Each chapter follows the same rhythm:

1. Introduction—a cold open roast of the physiologic failures you're about to meet.
2. Mechanism Breakdown—what's actually happening behind the symptoms.
3. Tables—written with snark but scientifically accurate.
4. Science Snapshot—exactly what you'd expect.
5. Translation—what the physiology *really* means in human terms.
6. Tilt Tip—actionable strategies with zero fluff.

This structure keeps the entire book cohesive, sharp, and readable—even when the physiology isn't.

Why Snark? Why Humor?

Because dysautonomia is brutal.

Because the physiology is relentless.

Because the testing is revealing.

Because the system itself is complicated, unforgiving, and frequently absurd.

Snark does not trivialize the condition; it humanizes it, reframes it, and keeps the reader awake long enough to understand it.

Science carries authority. Humor carries momentum. This book carries both—by force, if necessary.

The Point of No Return

Once you understand your autonomic system at this depth, you will never unsee its patterns.

You will recognize its logic everywhere.

You will understand the collapse before it happens.

You will spot the difference between compensation and catastrophe.

You will stop accepting explanations that rely on hand-waving and guesswork.

You will know the enemy. And you will know how to fight back. Now let's take control of the physiology.

PART I
KNOW YOUR ADVERSARY: THE AUTONOMIC SYSTEM UNMASKED

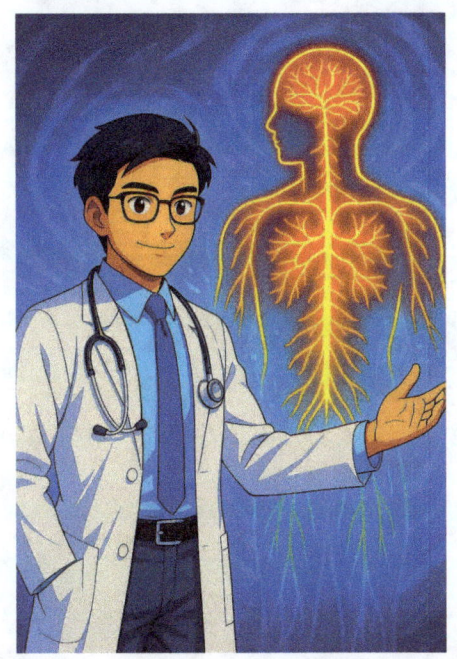

Where physiology stops pretending to be polite.

CHAPTER 1
THE AUTONOMIC NERVOUS SYSTEM: THE ORIGINAL MULTITASKER

Where the most overworked system in the body finally gets the respect, sarcasm, and diagnostic precision it deserves.

Overview: The System That Never Gets Credit Until It Implodes

Every medical specialty has its diva. Cardiology has its trophies. Neurology has its puzzles. Endocrinology has its algorithms.

But lurking quietly in the background—juggling cardiovascular control, temperature management, fluid distribution, digestion, inflammation, and about a dozen additional jobs no one wants—is the autonomic nervous system (ANS).

The ANS is the invisible general whose sole job is preventing you from passing out, overheating, starving, or spontaneously collapsing while you stroll around pretending you're self-maintaining.

It is relentless, indispensable, profoundly underappreciated, and the only system that receives attention only when it malfunctions so catastrophically that denial becomes impossible.

If the ANS worked in a hospital, it would be the night-shift charge nurse who runs the entire building while everyone else signs their notes, goes home, and takes credit.

This chapter establishes the battlefield: what the ANS controls, how it coordinates competing physiologic demands, what happens when it breaks, and why ignoring it is basically clinical denial packaged as confusion.

Welcome to the system behind the curtain.

Section 1—What the ANS Actually Controls (Spoiler: Everything That Matters)

Every organ system has a job. The autonomic nervous system holds the job description.

Here's the short list of what it micro-manages around the clock:

Physiologic Domains Under ANS Command

Domain	What the ANS Is Doing Every Second	What You Notice When It Fails
Blood pressure	Adjusting vascular tone to keep your brain perfused	Lightheadedness, "graying out," syncope
Heart rate & contractility	Modulating cardiac output to match demand	Tachycardia, palpitations, fatigue
Vascular resistance	Deciding where blood should be and when	Cold hands, heat intolerance, pooling
Thermoregulation	Coordinating sweating and skin blood flow	Heat collapse, chills, temperature swings
GI motility	Moving food without your permission or assistance	Bloating, nausea, slow transit
Bladder/bowel control	Reflexive sphincter coordination	Urgency, retention, "this isn't normal" moments

Domain	What the ANS Is Doing Every Second	What You Notice When It Fails
Endocrine cross-talk	Helping hormones hit their targets	Fatigue, instability, metabolic chaos
Pupillary reflexes	Adjusting to light and distance	Blurry vision, photophobia
Respiratory–cardiac coupling	Matching breath with beat	Dyspnea, air hunger, "I forget to breathe"
Immune modulation	Tuning inflammation	Flares, crashes after minor infections
Energy allocation	Deciding what gets fuel and what doesn't	Fatigue, post-exertional crashes

Lose just one domain? You get a tidy organ-based diagnosis. Lose several? Suddenly clinicians start whispering, *"Maybe anxiety?"* like it's a diagnostic code. In reality, multi-system symptoms are not dramatic. They're diagnostic.

Section 2—The Two-Branch System: The Dysfunctional Buddy-Cop Duo

Textbooks summarize autonomic control with the sophistication of a cartoon:

- Sympathetic = fight-or-flight
- Parasympathetic = rest-and-digest

Cute. Also simplistic to the point of being more mnemonic than meaningful.

The truth is that the ANS operates like a dysfunctional buddy-cop partnership—one officer is impulsive and caffeinated, and the other is exhausted but trying desperately to keep the peace.

The Sympathetic Nervous System (SNS)
Your accelerator.

A healthy SNS:
- Increases heart rate
- Maintains blood pressure
- Constricts blood vessels when you stand
- Redistributes blood to critical organs during stress
- Prevents you from face-planting during upright activities

When it misbehaves:
- It overfires → hyperadrenergic chaos
- It underfires → orthostatic hypotension
- It mistimes → delayed compensation and presyncope

In dysautonomia, standing upright becomes a full-body emergency drill. Your body behaves like there's an ill-tempered lion lurking in every hallway.

The Parasympathetic Nervous System (PNS)
Your braking system.

A healthy PNS:
- Dampens inflammation
- Slows heart rate
- Guides digestion
- Regulates HRV (heart rate variability)
- Promotes recovery and repair

When impaired:
- HRV collapses
- Tachycardia becomes baseline
- Sleep stops restoring
- Digestion becomes optional
- Everything feels like "too much"

The PNS isn't gentle. It's outnumbered.

Section 3—The Command Center No One Talks About: The Brainstem

If the ANS is an army, the brainstem is command headquarters. Specifically, the nucleus tractus solitarius, dorsal vagal complex, ventrolateral medulla, and pontine centers.

These structures coordinate blood pressure, breathing, heart rate, reflex arcs, and every "automatic" function people mistakenly believe is effortless.

When brainstem signaling falters, reflexes misfire, timing becomes unstable, blood pressure fluctuates, recovery slows, and symptoms become wildly unpredictable.

Most clinicians treat the brainstem as background scenery. Autonomic medicine treats it as the plot.

Section 4—How the ANS Manages Competing Priorities (Without Applause)

Every moment of your life, the autonomic nervous system (ANS) is juggling contradictions—keeping the brain perfused, maintaining blood pressure, avoiding overheating, digesting food, regulating immune activity, preventing fainting during upright motion, conserving energy, and mobilizing energy.

These demands overlap, conflict, and require rapid prioritization.

Autonomic Triage Table

Situation	What the ANS Should Do	Where It Fails in Dysautonomia
Standing up	Increase HR, constrict vessels, stabilize BP	HR skyrockets; BP collapses; brain perfusion wobbles
Eating	Redirect blood to the gut, maintain perfusion	Post-meal fatigue, tachycardia, near-syncope
Heat exposure	Increase sweat, dilate skin vessels	Overheating, blood pooling, tachycardia
Stress	Sympathetic surge with fast recovery	Prolonged surges, adrenaline dumps, delayed recovery
Exercise	Increase output, maintain tone	Post-exertional crashes, prolonged tachycardia
Sleep	Vagal dominance, physiologic reset	Fragmented sleep, poor recovery, morning crashes

This is not a subtle disease. This is an infrastructure failure.

Section 5—What Happens When the ANS Breaks (The Real Explanation Behind "Mystery Symptoms")

When the ANS falters, everything downstream improvises badly.

Symptoms emerge across multiple systems, including cardiovascular, gastrointestinal, temperature regulation, respiratory, cognitive, musculoskeletal, sensory, and immune domains.

Patients don't get "weird symptoms." They get predictable consequences of impaired coordination.

Mechanism → Failure → Real-World Manifestation

Failure Mode	Physiologic Breakdown	What Patients Experience
Vagal weakness	Poor HR modulation	Tachycardia, poor recovery, exertional crashes
Sympathetic underactivation	Inadequate vasoconstriction	Orthostatic hypotension, fainting
Sympathetic overactivation	Excessive compensation	Hyperadrenergic POTS, tremors
Baroreflex instability	Delayed correction	Dizziness, blood pressure volatility
Small fiber neuropathy	Impaired sweat/skin signaling	Heat intolerance, temperature swings
Central dysautonomia	Timing and coordination failure	Unpredictable crashes, extreme variability

This is not "sensitivity." This is physiology confessing its failures.

Section 6—Why Testing the ANS Isn't Optional

Autonomic dysfunction hides exceptionally well at rest—and spectacularly poorly under duress.

Testing matters because autonomic symptoms are nonspecific, pattern recognition differentiates mechanisms, compensation can mask failure, real-world physiology rarely resembles clinic physiology, and symptoms without context mislead clinicians.

Without autonomic testing, you are diagnosing in the dark.

Why Clinicians Often Miss It

Reason	Translation
Vitals look normal.	They only measured two numbers while you were seated.
Symptoms are across multiple systems.	Exactly. That's the point.
Maybe it's stress.	Stress worsens ANS dysfunction; it doesn't cause it.
Try more hydration.	Advice appropriate for a cactus, not a patient.
Everything looks normal.	The testing wasn't.

Autonomic testing is the difference between guessing and knowing.

Section 7—Why Understanding the ANS Is the Key to Everything Else

This book—every chapter, every mechanism, every treatment strategy—rests on a single foundation: If you don't understand the autonomic system, you cannot understand dysautonomia.

Every downstream issue ties back to reflex function, timing, coordination, reserve, and failure mode. The ANS is not a subsystem—it is the operating system.

Translation—What This Chapter Really Means

- Your ANS is running the entire show, no matter what organ a clinician decides to blame.
- Symptoms across multiple systems aren't theatrical—they're diagnostic.
- The sympathetic and parasympathetic systems are constantly negotiating survival behind the scenes.
- If the brainstem misfires, everything else does too.
- "Normal vitals" mean nothing without context, posture, timing, or stress.
- If clinicians ignore the ANS, that's not your fault. That's the gap in their training.

Tilt Tip—How to Use This Knowledge Right Now

- When telling your story to clinicians, emphasize patterns, not individual symptoms.
- When tracking symptoms, categorize by mechanism (heat, posture, meals, exertion, stress).
- If someone says, "It's anxiety," ask them to explain your HRV curve, your Valsalva pattern, and your standing pulse pressure.
- Use this chapter's tables to anchor your understanding of which branch of the ANS is unplugging itself when life gets interesting.
- And above all—remember that nothing about your symptoms is random. They are the predictable consequences of a system fighting to keep you upright.

CHAPTER 2
HEART RATE VARIABILITY: LISTENING TO THE HEART–BRAIN CONVERSATION (EVEN WHEN THEY'RE ARGUING)

Where HRV stops being a wellness fashion accessory and becomes the earliest confession your autonomic system ever makes.

Introduction: The Most Misunderstood Autonomic Signal in Medicine

Heart Rate Variability (HRV) is one of the most powerful physiologic metrics ever discovered—and one of the least understood.

The world treats HRV like a meditation badge, a sleep score accessory, a personality test for your vagus nerve, or worse, a lifestyle flex on social media.

Meanwhile, those of us in autonomic medicine treat HRV for what it actually is: raw telemetry from the command chain of survival.

HRV doesn't care about your mood, your meditation exercises, or whether you downloaded a breathwork app on sale. HRV is the earliest signal of whether your autonomic system is flexible or rigid, adaptable or exhausted, prepared or failing, compensating or collapsing.

This chapter teaches you how to interpret HRV as a clinical weapon, not a vibe.

Section 1—What HRV Actually Measures: Spoiler, Not "Calm"

HRV measures the millisecond variation between consecutive heartbeats—the RR intervals.

Healthy physiology produces dynamic variation, moment-to-moment adaptation, and subtle negotiation between autonomic branches.

Low HRV means one thing: the system has lost its ability to improvise.

A rigid, unchanging heart rhythm under normal conditions is not "disciplined physiology." It's a sign that the nervous system is petrified.

To understand how revealing RR intervals are, compare how a healthy and unhealthy system responds to the smallest stressors:

HRV Under Challenge: Side-by-Side Reality Check

Stressor	Healthy HRV	Low HRV
Standing up	HR fluctuates briefly, then stabilizes	HR becomes linear panic
Mild heat	Brief dip in parasympathetic tone, then recovery	Sympathetic surge with no recovery
Light exercise	Predictable rise and fall with cooldown	HR stays elevated for hours
Cognitive load	Subtle pattern changes	Flatline unpredictability
Emotional stress	Flexibility → recovery	Loss of modulation → crash

If RR intervals stop dancing, adaptability is gone—and adaptability is the only currency that matters in autonomic medicine.

Section 2—RR Intervals: The Real Story Behind the Pulse

Your heartbeat is not supposed to behave like a metronome. If it does, your autonomic system is terrified of improvisation.

Each RR interval reflects three interacting forces: sympathetic activation, parasympathetic withdrawal, and baroreflex modulation.

These micro-adjustments allow your body to maintain blood pressure and perfusion during constantly shifting conditions.

The RR Interval Breakdown

Component	Physiologic Role	What Happens When It Fails
Sympathetic drive	Increases HR, vasoconstriction	Flat response, delayed compensation
Parasympathetic tone	Slows HR, modulates variability	Rigid intervals, low RMSSD*
Baroreflex feedback	Balances each beat	Pattern mismatch, instability

Root Mean Square of the Successive Differences—one of the most widely used heart-rate variability (HRV) metrics.

A smooth pulse on a smartwatch looks comforting but tells you nothing about autonomic truth. The variability inside those beats is where the confession lives.

Section 3—The Two Domains of HRV: Time vs. Frequency

Two different ways to catch the system in a lie.

A. Time Domain Metrics—The Big Picture

These metrics capture overall variability:

- SDNN—Standard Deviation of NN intervals (global autonomic variability)
- RMSSD—Root Mean Square of Successive Differences (parasympathetic integrity)

How to interpret time domain values

Pattern	Meaning	Clinical Interpretation
Low SDNN	Low global variability	Depleted autonomic reserve
Low RMSSD	Low short-term variability	Vagal impairment
Both low	Flatline variability	Global dysautonomia

Time domain metrics are the HRV equivalent of checking someone's bank account: Is there reserve? Is there flexibility? Is there anything left to spend?

B. Frequency Domain Metrics—The Truth Serum

Frequency analysis breaks HRV into electrical signature components:

- HF (High Frequency) → parasympathetic tone
- LF (Low Frequency) → baroreflex + sympathetic/parasympathetic interplay

- LF/HF ratio → balance between accelerator and brake
- Total Power → overall autonomic energy supply

Frequency Domain Insights

Metric	Physiologic Meaning	Interpretation Error to Avoid
HF	Vagal tone	Low HF ≠ stress; it = vagal weakness
LF	Baroreflex + mixed ANS	Not "sympathetic only"
LF/HF	Balance	High ≠ anxiety; may = compensation
Total Power	Autonomic reserve	Low = fragile physiology

If time domain is the bank account, frequency domain is the spending habit analysis.

Section 4—When HRV Doesn't Just Decline—It Diagnoses

Low HRV is not a mood. It is a mechanism.

Diagnostic Patterns That Actually Matter

Pattern	Mechanism	What It Suggests
Low RMSSD + normal LF	Vagal impairment	Early parasympathetic failure
Normal HF + low LF	Baroreflex weakness	Impaired reflex coupling
Both low	Global deficit	Multifocal dysautonomia
High LF/HF	Sympathetic dominance	Hyperadrenergic states, illness

Pattern	Mechanism	What It Suggests
High HF + low HR	Athletic vagal phenotype	Not common in dysautonomia

HRV doesn't whisper. It reveals the entire story before other tests even begin.

Section 5—HRV Under Stress: Recovery Is the Real Examination

Baseline HRV is useful. But recovery is where the system either proves itself or confesses collapse.

In healthy physiology, HR rises, HF drops, LF rises, and everything normalizes quickly.

In dysautonomia, HR skyrockets, HF stays suppressed, LF never stabilizes, and recovery becomes mythical.

This is why patients wake up exhausted after "eight hours of sleep"—their autonomic system never turned off the alarms.

Recovery Curve Differences

Feature	Healthy System	Dysautonomia
Peak HR	Brief	Excessive
HF suppression	Temporary	Persistent
LF response	Robust	Delayed or absent
Return to baseline	Minutes	Hours or never
Perception	"That was easy."	"Why do I feel wrecked?"

Autonomic injury announces itself in its inability to recover.

Section 6—Medications: The Great HRV Impersonators

HRV interpretation is nearly impossible without understanding what medications are doing in the background.

How common meds distort HRV

Medication Class	What It Distorts	Resulting False Impression
Beta-blockers	HR variability, sympathetic tone	False vagal strength, hidden POTS
Anticholinergics	HF + RMSSD	Fake vagal failure
SNRIs	LF/HF	False sympathetic dominance
Stimulants	LF + HR	Hyperadrenergic mimicry
Benzodiazepines	HF	Cosmetic parasympathetic bump

A normal HRV on medication is not inherently normal. It may be cosmetic.

Section 7—Why HRV Should Be the First Autonomic Test Performed

HRV screens: adaptability, stress load, autonomic reserve, baroreflex performance, and vagal tone.

Low HRV guarantees poor orthostatic performance, reduced exercise tolerance, impaired thermoregulation, poor recovery, and vulnerable physiology.

If HRV is abnormal, nothing else in the autonomic test suite will surprise you. If HRV is normal but symptoms are severe? Be suspicious. Someone is compensating—poorly.

Section 8—Science Snapshot: HRV, the Early Warning Radar

The Hierarchy of HRV Interpretation

Level	Question Asked	What It Reveals
1	Is variability present?	Basic flexibility vs. rigidity
2	Which domain is impaired?	Parasympathetic vs. baroreflex vs. global
3	Does stress exaggerate weakness?	Mechanistic vulnerability
4	Does recovery occur?	Reserve and prognosis
5	Are meds masking findings?	Artifact vs. physiology

This snapshot alone can classify 70% of dysautonomia phenotypes before tilt/stand testing even begins.

Translation—What This Chapter Really Means

- HRV isn't a relaxation metric. It's a survival metric.
- Low HRV means the system has stopped improvising and is running in emergency mode.
- Frequency domain analysis is not optional; it's diagnostic clarity.
- Medications distort HRV so aggressively that interpreting it cold is a clinical gamble.
- Recovery patterns matter more than resting patterns.
- If HRV collapses, everything else in the autonomic system will eventually follow.

Tilt Tip—How to Use HRV in Real Life

- Record HRV during normal life, not during your "perfect calm day."

- A single low HRV day means nothing; a pattern means everything.
- Compare HRV before and after triggers: heat, meals, stress, and upright posture.
- If HRV stays suppressed after exertion, plan recovery—the system will not self-correct.
- When clinicians doubt symptoms, HRV data becomes evidence, not emotion.

CHAPTER 3
BREATHING, BEARING DOWN, AND STANDING UP: THE CORE AUTONOMIC REFLEXES

Because nothing exposes autonomic dysfunction faster than asking the body to do the three things it was supposedly built to handle.

Introduction: The Reflexes That Decide Whether You Stay Upright

If HRV reveals the autonomic system's adaptability, these reflex tests reveal its competence—and its secrets.

Deep breathing. Valsalva. Standing upright.

Three basic physiologic tasks humans should be able to perform without summoning paramedics, bystanders, or divine intervention.

Yet for people with autonomic dysfunction, these three "simple" tasks reliably expose timing failures, vascular incompetence, parasympathetic collapse, sympathetic overreaction, baroreflex confusion, and the nervous system's overall ability to negotiate with gravity.

These reflexes aren't wellness rituals or breathing exercises. They are interrogations.

They ask the autonomic system:

1. Can you modulate?
2. Can you compensate?
3. Can you recover?
4. And can you do any of this without fainting, panicking, overheating, or summoning an ambulance?

Consider this chapter your introduction to the three tasks that separate stable physiology from chaos.

Section 1—Deep Breathing: The Parasympathetic Stress Test for a System That Hates Stress

Deep breathing is often marketed as mindfulness, relaxation, a way to "center yourself," or a spiritual tune-up recommended by people who have never fainted in a grocery store.

In autonomic medicine, deep breathing is far less poetic. It is a vagal interrogation. It asks, "Parasympathetic system, are you actually awake?"

What Healthy Looks Like

A functioning parasympathetic system generates a smooth, elegant sinusoidal pattern.

Phase	Expected Response	Meaning
Inhale	HR rises (vagal withdrawal)	The brake releases cleanly
Exhale	HR falls (vagal engagement)	Strong, immediate braking
Oscillation	15+ bpm amplitude	Parasympathetic system is competent
Pattern	Smooth waves	No sympathetic contamination

When the curve looks like art, the ANS is doing its job.

What Dysfunction Looks Like
Impaired deep breathing is where things become diagnostic gold.

Finding	Physiologic Problem	Clinical Meaning
Shallow oscillation	Weak vagal activation	Early parasympathetic decline
Delayed recovery	Baroreflex lag	Fragile modulation
Rigid, linear HR	Sympathetic contamination	Overdrive physiology
Flatline response	Parasympathetic failure	Severe autonomic impairment

Low respiratory-driven variability is not anxiety. It is not "poor breathing technique." It is parasympathetic insufficiency—overwhelmed or injured.

Why It Matters
Parasympathetic impairment predicts exaggerated upright HR surges, slow recovery from stress, post-exertional exhaustion, temperature intolerance, digestive dysfunction, and non-restorative sleep.

If the brake doesn't work while seated, it will absolutely fail the moment gravity enters the scene.

Section 2—The Valsalva Maneuver: A Sympathetic Stress Test Without a Treadmill
The Valsalva is not meditation. It is not breathwork. It is not a "technique." It is a controlled internal crisis—twenty seconds of physiologic honesty.

This test forces the autonomic system to reveal sympathetic activation strength, vascular competence, baroreflex timing, parasympathetic braking, and coordination across all branches.

Deep breathing tests the brake. Valsalva tests the entire vehicle—on ice—during a storm.

The Four Phases That Make Guessing Impossible

Each phase exposes a different piece of autonomic machinery.

Phase I—Mechanical Spike

- You strain (i.e., bear down or breathe into a tube).
- Intrathoracic pressure rises.
- BP increases briefly.

No nerves involved—just physics.

Phase II—The Drop and the Fight

- Venous return falls.
- BP drops.

The sympathetic nervous system should counterattack:

- HR rises
- Vascular tone increases
- BP stabilizes

Phase II is the sympathetic midterm exam.

Phase III—Release

- Strain stops.
- BP dips again.
- Compensation should already be in motion.

A delayed correction here means trouble.

Phase IV—Overshoot and Brake

- BP overshoots baseline.
- Parasympathetic braking (vagal response) kicks in.
- HR falls sharply.

Phase IV is the parasympathetic system's report card.

What Healthy Looks Like

Element	Expected Pattern	Meaning
Phase II HR rise	Rapid	Intact sympathetic response
Phase II BP stabilization	Strong	Competent vasoconstriction
Phase IV overshoot	Crisp	Baroreflex + vascular strength
HR braking	Immediate	Strong vagal tone

These responses indicate elegance, coordination, and physiologic resilience.

What Dysfunction Looks Like

Abnormality	Failure Mode	Interpretation
Weak Phase II HR rise	Sympathetic denervation	Neuropathic disorders, central failure
Huge HR spike but stable BP	Hyperadrenergic physiology	Excessive norepinephrine
Absent Phase IV overshoot	Baroreflex failure	Timing and coordination breakdown

Abnormality	Failure Mode	Interpretation
Slow recovery	Mixed dysfunction	Post-viral, autoimmune, progressive neuropathies
BP collapse after release	Vascular incompetence	Common in fainters

Valsalva abnormalities are not vague—they are specific fingerprints of failure.

Chronotropic Competence vs. Chronotropic Desperation
Clinicians often declare, "Your heart rate increased appropriately." Without Valsalva, this statement is guesswork wearing a badge.

Chronotropic competence → HR rises because the system is healthy.

Chronotropic desperation → HR rises because the blood vessels refuse to constrict, and the heart panics to compensate.

Same HR curve. Opposite mechanism. Opposite treatments.

Section 3—Standing Up: Gravity's Final Exam
Standing upright is the most underrated physiologic challenge humans perform. Every other test in autonomic medicine is a simulation. Standing is the real world.

Upon standing:
- Blood pools
- Venous return drops
- Stroke volume falls
- Cerebral perfusion becomes negotiable
- Sympathetic activation must save the day

A healthy autonomic system manages all of this within 5 seconds.

Healthy Orthostatic Response

Timeframe	Expected Action	Meaning
0–5 seconds	SNS fires, HR rises 10–20 bpm, BP stabilizes	Intact reflex
30 seconds	Tone adjusts, perfusion normalizes	System is coordinated
2 minutes	Steady state; no symptoms	Physiologic perfection

Anything else is not a personality trait or overreaction. It is physiology under siege.

Orthostatic Failure Patterns
1. POTS (Postural Orthostatic Tachycardia Syndrome)
    - HR increase ≥30 bpm (or HR >120)
    - No hypotension
    - Compensation, not anxiety
    - Survival strategy, not fragility

2. Orthostatic Hypotension
    - BP falls significantly
    - Insufficient sympathetic activation
    - Vascular failure → cerebral perfusion collapse

3. Neurocardiogenic Syncope
    - Initial compensation
    - Sudden vagal surge
    - HR and BP drop
    - Instant collapse

4. Mixed Orthostatic Intolerance
- HR spikes
- BP wobbles
- Venous pooling
- Delayed recovery

A negotiation with gravity—one the patient keeps losing.

Standing Still: The Silent Killer of Orthostatic Reserve
Movement pumps blood. Stillness does not. Standing still increases pooling, decreases stroke volume, and stresses the system more than walking.

Symptoms include dimming vision, nausea, heat rising, tremors, tachycardia, and forward collapse.

This is not deconditioning. It is hemodynamic physics.

Section 4—Why These Three Reflexes Matter More Than Any Others

These three tasks examine the core responsibilities of the autonomic nervous system:

1. Deep Breathing (Parasympathetic Function):
- Timing
- Amplitude
- Vagal integrity

2. Valsalva (Sympathetic + Baroreflex Function):
- Compensation
- Coordination
- Vascular competence

3. Standing (Integrated Real-World Performance):
   - Circulation
   - Perfusion
   - Survival under gravity

Reflex Test Hierarchy Table

Test	What It Measures	Why It Matters
Deep Breathing	Parasympathetic modulation	Earliest sign of failure
Valsalva	Sympathetic + baroreflex	Diagnostic fingerprints
Standing	Integrated function	Real-world correlation

When a system fails these reflexes, it will not tolerate heat, meals, exercise, stress, prolonged upright posture, and inconsistent environments.

Reflex testing is not academic—it is the foundation of clinical clarity.

Science Snapshot: The Reflex Arc Logic

Three Tests → Four Clues → One Mechanism

Domain	Deep Breathing	Valsalva	Standing	Interpretation
Para-sympathetic	Amplitude	Phase IV braking	Recovery	Vagal integrity
Sympathetic	Contamination	Phase II response	Initial HR rise	Activation strength
Baroreflex	Timing	BP & HR transitions	Pulse pressure	Coordination
Vascular	None directly	BP stabilization	BP stability	Tone and competence

This is how mechanisms differentiate themselves. Not by symptoms—but by reflex behavior.

Translation—What This Chapter Really Means

- If deep breathing is impaired, the brake is failing.
- If Valsalva is impaired, the wiring or timing is failing.
- If standing is impaired, the whole system is failing.
- These tests aren't optional—they diagnose the mechanism behind the symptoms.
- "But your vitals are normal" is not an interpretation; it's an oversight.
- These three reflexes will reveal dysfunction long before clinic vitals ever will.

Tilt Tip—How to Use This Knowledge Right Now

- Pay attention to which reflex triggers symptoms: breathing, bearing down, or standing.
- If you crash after standing but perform well seated, your problem is likely circulatory, not emotional.
- If Valsalva knocks you out, ask your clinician to evaluate sympathetic activation and baroreflex timing.
- If deep breathing is impaired, every other stressor will exaggerate symptoms.
- Use this chapter to track patterns: which mechanism collapses first, how fast, and under what conditions.

CHAPTER 4
WHEN THE SKIN TALKS: SUDOMOTOR AND SMALL FIBER FUNCTION

Because sweat glands tell the truth long before anything else does.

Introduction—The Most Honest System in the Body

If the autonomic nervous system is the shadow government of human physiology, the skin is its whistleblower—loud, blunt, and absolutely unconcerned with protecting anyone's feelings.

Sudomotor function reveals autonomic dysfunction the way crime-scene footprints reveal suspects: clearly, directly, and without room for the forensic fan fiction some call "clinical judgment."

Sweat glands don't negotiate. They don't "try their best." They don't fake a performance for your clinician's approval. They either work—or they don't.

When sweat output declines, delays, or disappears, there is exactly one physiologic explanation: neurologic dysfunction.

Not hydration status. Not mood. Not "being sensitive to heat." Not dramatic flair.

Sudomotor failure is the earliest sign that small autonomic fibers are struggling. And once these fibers falter, the rest of the system will inevitably follow.

This chapter explains why sweat is the most honest biomarker in autonomic medicine and why testing it turns vague symptoms into precise maps.

Section 1—The C-Fibers: The Autonomic Canary in the Coal Mine

Sudomotor nerves are unmyelinated C-fibers—tiny, slow, metabolically fragile, and the first to fail when anything in the body goes wrong.

These fibers regulate sweat gland activation, skin blood flow, temperature management, microvascular distribution, sensory signaling, and neuroimmune communication.

They are so finely tuned that nearly every systemic insult—metabolic disease, autoimmunity, toxic exposure, nutritional deficiency, post-viral injury—hits them first.

C-Fiber Dysfunction: Mechanism → Consequence

C-Fiber Role	What Failure Looks Like	How Patients Experience It
Sweat activation	Reduced sweat output	Overheating, intolerance of heat, shower collapse
Thermoregulation	Inadequate heat shedding	Red skin, blotchy skin, temperature swings
Cutaneous vasodilation	Poor shunting of blood	Pooling, cold hands, blood redistribution issues
Microvascular flow	Uneven tissue perfusion	Exercise crashes, leg heaviness

C-Fiber Role	What Failure Looks Like	How Patients Experience It
Sensory signaling	Altered perception	Burning/tingling, pain, numbness
Neuroimmune link	Dysregulated inflammation	Post-viral worsening, flare reactivity

Small fiber impairment is not a footnote—it's the prequel to systemic autonomic instability.

Section 2—Electrochemical Skin Conductance: The Non-Invasive Biopsy

Electrochemical Skin Conductance (ESC) evaluates sweat gland function by measuring chloride ion transport under controlled stimulation. It checks whether your sweat glands still have a functioning nerve supply—without needles, incisions, or invasive procedures.

ESC Interpretation Table

Result	Meaning	Clinical Interpretation
High conductance	Intact innervation	Healthy small fibers
Low conductance	Impaired innervation	Neuropathy—peripheral autonomic injury
Borderline	Early degeneration	Small fiber compromise beginning
Asymmetry	Localized nerve injury	Trauma, radiculopathy, regional problem
Hand + foot decline	Systemic neuropathy	Metabolic, autoimmune, post-viral processes

Because feet have the longest fibers, they fail first. Hands fail later, when disease becomes systemic. Think of ESC as a non-invasive biopsy of nerve survival—or a polygraph test the ANS can't cheat on.

Section 3—Patterns That Matter: Sudomotor Maps of Disease

Sudomotor patterns don't require guesswork. They diagnose themselves.

Pattern 1—Low Feet, Normal Hands: Early Length-Dependent Neuropathy

Common causes:
- Diabetes
- Pre-diabetes/Insulin resistance
- Metabolic syndrome
- Nutritional deficiencies
- Early autoimmune disease
- Post-viral injury

This is the earliest sign that small fibers are "retiring" at the farthest end of the network.

Pattern 2—Low Feet + Low Hands: Global Small Fiber Neuropathy

Seen in:
- Sjögren's
- Lupus
- Autoimmune ganglionopathy
- Long COVID
- Diabetic neuropathy
- Amyloidosis

This is not local dysfunction. It's a systemic breakdown of small fiber integrity.

Pattern 3—Patchy or Asymmetric Loss: Localized Lesions
Consider:
- Radiculopathy
- Spinal pathology
- Trauma
- Entrapment neuropathy

Localization tells you exactly where the problem is—better than imaging half the time.

Pattern 4—Normal Sweat + Severe Symptoms: Central Dysautonomia
When sudomotor fibers are intact but orthostatic symptoms, heat intolerance, baroreflex weakness, cognitive fog, and tachycardia are severe, the culprit is central—the command center, not the wires.

Pattern 5—Declining ESC Over Time: Progressive Neuropathy
The earliest warning that disease is progressing—not stabilizing. This is where follow-up matters.

Section 4—Heat Intolerance: A Predictable Physiologic Catastrophe

Heat is a diagnostic wrecking ball for the autonomic system. It simultaneously tests:
- Vasodilation
- Sweat gland responsiveness
- Cardiac output
- Blood pressure regulation
- Cerebral perfusion

Why Heat Overwhelms Dysautonomia Patients

Step	Physiologic Breakdown	Symptom
1	Massive skin vasodilation	Lightheadedness
2	Blood shifts away from the core	Tachycardia
3	Decreased venous return	Pooling, dizziness
4	Decreased cerebral perfusion	Fatigue, presyncope
5	Impaired sweat response	Overheating
6	SNS overdrive	Tremors, flushing

Heat intolerance is a small fiber problem long before it becomes a blood pressure problem. This is not emotional sensitivity to temperature. This is an engineering failure.

Section 5—When Cardiac Tests Look Normal but Life Doesn't

Some patients "pass" HRV, deep breathing, Valsalva, and orthostatic vitals. Yet daily life is a disaster. Why? Because central reflexes can compensate only so long before peripheral fibers give out.

These patients typically experience heat-triggered crashes, shower-induced near-syncope, exercise intolerance, burning pain, foot numbness, and temperature instability.

Sudomotor testing explains everything. The wires are failing even if the command center is putting on a good show.

Section 6—Why Sudomotor Dysfunction Happens

Sudomotor decline is never random. It always traces back to a cause.

1. Metabolic Causes:
   - Diabetes
   - Pre-diabetes
   - Dyslipidemia
   - B12 deficiency
   - Folate deficiency
   - Copper deficiency

2. Autoimmune/Post-Infectious Causes:
   - Sjögren's
   - Lupus
   - Autoimmune autonomic ganglionopathy
   - Celiac-related neuropathy
   - Post-COVID dysautonomia

3. Neurodegenerative Causes:
   - Amyloidosis
   - Parkinsonian autonomic disorders
   - Hereditary neuropathies

The unifying theme: C-fibers hate metabolic stress, immune dysfunction, and toxins. They falter early and loudly.

Section 7—How Sudomotor Dysfunction Shapes Real-World Physiology

Sweat loss is not a cosmetic inconvenience. It destabilizes everything else.

When Sweat Fails → Everything Follows

Failure	Downstream Issue	Real-World Impact
Heat shedding fails	Body overheats rapidly	Shower collapses, sauna intolerance
Skin blood flow increases	Blood pools in extremities	Orthostatic symptoms
Venous return drops	Cardiac output decreases	Fatigue, "I have to sit down now"
Cerebral perfusion declines	Cognitive fog	Confusion, slow processing
Microvascular tone falters	Tissue perfusion losses	Exercise crash, heavy legs

Small fibers may be tiny, but their collapse causes system-wide chaos.

Science Snapshot: Sudomotor Truth Table

Small Fiber Testing → Mechanism → Treatment Direction

Sudomotor Finding	Mechanism	Implication
Low feet	Early neuropathy	Metabolic workup, early intervention
Low feet + hands	Systemic neuropathy	Autoimmune/Inflammatory evaluation
Asymmetric	Regional nerve injury	Imaging or nerve-specific evaluation
Normal ESC + severe symptoms	Central dysfunction	Baroreflex, brainstem-focused testing

Sudomotor Finding	Mechanism	Implication
Decline over time	Progression	Repeat testing, aggressive risk modification

Sudomotor data anchors the entire diagnostic framework.

Translation—What This Chapter Really Means

- Sweat glands are your early-warning alarms—not optional side characters.
- Sudomotor decline is the earliest sign that the autonomic system is losing ground.
- Heat intolerance is not drama—it is predictable physics in a failing system.
- Normal sweat results with severe symptoms mean the problem is central, not peripheral.
- Small fiber loss explains why many patients "look normal" in clinic but collapse at home.
- These patterns aren't vague—they are diagnostic fingerprints.

Tilt Tip—How to Use This Knowledge Right Now

- Pay attention to heat responses—they are diagnostic gold.
- Track whether symptoms start in the feet, hands, or both—that localizes neuropathy.
- If your sweat test is normal but symptoms are severe, push for baroreflex and central testing.
- If ESC declines over time, treat it as a sign to escalate evaluation—not reassurance.
- Use symptom triggers (heat, showers, exercise) as data, not "overreactions."

PART II
THE WEAPONS ROOM: HOW WE INTERROGATE YOUR NERVOUS SYSTEM

Where testing stops being optional and becomes diagnostic warfare.

CHAPTER 5
RESTING STATE:
THE AUTONOMIC BASELINE

Because if your body can't maintain stability while doing nothing, it's definitely not going to hold together once gravity or life events get involved.

Introduction—The Quietest Moment That Reveals the Loudest Truth

Every great interrogation starts the same way: "Let's begin with some simple questions."

In autonomic medicine, the "simple questions" are asked while the patient is sitting or lying down, breathing normally, in a cool room, and doing absolutely nothing. This moment—your resting baseline—is where the autonomic system reveals its default setting.

A healthy nervous system at rest is steady, adaptable, relaxed, responsive, and balanced.

A dysfunctional system at rest is compensating, unstable, guarded, over-alert, and burning resources it does not have.

And here's the part people miss: If the ANS can't maintain homeostasis at rest, it will absolutely collapse under challenge.

Resting state is not a warm-up. It is the foundation upon which every autonomic reflex sits. If this foundation is cracked, the building will shake during every test, every stressor, and every upright moment.

This chapter teaches you how to read the baseline like a confession.

Section 1—What the Resting State Should Look Like

A physiologically normal baseline shows:
- Normal heart rate
- Normal blood pressure
- High HRV
- Balanced sympathetic/parasympathetic tone
- Crisp respiratory coupling
- Predictable sinus rhythm

The resting autonomic system should be so steady that clinicians rarely think about it—because when it works, it is invisible.

Healthy Resting Physiology: The Ideal

Parameter	Healthy Range	Meaning
HR	Typically 60–85 bpm*	Efficient cardiac output
BP	Stable, predictable	Competent vascular tone
HRV	High RMSSD, stable SDNN	Robust vagal reserve
Respiratory sinus arrhythmia	Present	Strong parasympathetic brakes
Variability	Smooth	Adaptable and flexible

If 100 bpm is the ceiling, 60–85 is the sweet spot. Everything in between is negotiable.

A healthy baseline reflects a system that is ready for posture changes, cognitive load, temperature variation, exercise, digestion, stress, and anything resembling daily life.

The resting state is the energy reserve your autonomic system uses to survive existence.

Section 2—What the Resting State Should NOT Look Like

Resting physiology becomes pathologic when:
- HR is excessively high
- HR is tightly fixed and unchanging
- BP wobbles for no reason
- HRV is depressed
- Respiratory coupling is minimal
- Sympathetic "noise" is detectable at rest

These abnormalities are not subtle. They're just frequently ignored.

Pathologic Baseline Patterns

Finding	Mechanism	Clinical Meaning
Resting tachycardia	Sympathetic dominance	Low stroke volume, poor reserve
Fixed HR	Loss of variability	Vagal weakness, baroreflex impairment
Low HRV	Autonomic exhaustion	Poor recovery capacity
Labile BP	Vascular instability	Impaired sympathetic coordination
Rapid shallow breathing	Compensatory	Poor vagal modulation

Most autonomic failure patterns are already obvious before you stand up, breathe deeply, or attempt Valsalva. Resting baseline is the prequel—and it always contains spoilers.

Section 3—The Physiology Behind the Baseline: Why "Doing Nothing" Is Actually Complex

The resting state is not easy for the ANS. It requires an ongoing balance between metabolic demand, baroreflex stabilization, vagal parasympathetic engagement, sympathetic quieting, smooth respiratory coupling, thermal regulation, and background endocrine communication.

In other words, resting involves more physiologic coordination than people realize.

Why Rest Requires Precision

System	Baseline Task	What Goes Wrong in Dysautonomia
Sympathetic	Remain suppressed	Overactivation, noise, tremulous HR
Parasympathetic	Remain dominant	Vagal failure, rigidity
Baroreflex	Stabilize pressure	Delayed response, BP wobble
Circulation	Maintain venous return	Pooling even when seated
CNS (central nervous system)	Modulate tone	Hypervigilance, instability

The ANS must use minimal energy while maintaining maximal stability. This is a surprisingly difficult job—and the first job to fail.

Section 4—The Baseline HR: A Window Into Compensatory Physiology

Resting heart rate in autonomic medicine is not about fitness—it's about compensation.

1. Resting Tachycardia: The early red flag
Resting HR is high because:
- Stroke volume is low
- Blood pooling has already begun
- Sympathetic "pre-activation" is compensating
- Parasympathetic reserve is depleted
- The body is fighting gravity before gravity even arrives

Why Resting Tachycardia Isn't Anxiety

Myth	Reality
"Your pulse is fast because you're nervous."	Anxiety cannot drop stroke volume.
"This is just stress."	Stress cannot reduce venous return while supine.
"Maybe you're just out of shape."	Deconditioning doesn't cause sympathetic overdrive at rest.

Resting tachycardia is physiology—not psychology.

2. Resting Bradycardia: When it is meaningful and when it is not
Athletes:
- High vagal tone
- High HRV
- Normal respiratory coupling

Dysautonomia patients:
- Sometimes low heart rate
- But paired with low HRV and abnormal Valsalva

One is a sports adaptation. The other is a wiring malfunction.

3. The "Fake Normal" HR

Many patients present with a normal resting HR that is quietly masking:
- Low HRV
- High LF/HF ratio
- Early parasympathetic decline
- Latent sympathetic overdrive

Always compare resting HR with HRV. The combination—not the number —reveals truth.

Section 5—Resting Blood Pressure: Stability or Hidden Chaos

BP at rest is supposed to be boring. Boring is good.

When BP is *not* boring—when it wanders, dips, or spikes during rest—it indicates:
- Vascular incompetence
- Baroreflex dysfunction
- Central instability
- Sympathetic misfiring

Resting BP Patterns That Matter

Pattern	Mechanism	Interpretation
Wide pulse pressure	Excessive vasodilation	Vascular under-activation
Narrow pulse pressure	Poor stroke volume	Orthostatic intolerance risk
Labile BP	Baroreflex timing issues	Unreliable compensation
High-normal BP	Sympathetic baseline	Overdrive physiology
Low-normal BP	Low peripheral resistance	Easy presyncope trigger

Clinicians often ignore mild BP irregularities at rest, not realizing they predict upright collapse with shocking accuracy.

Section 6—Respiratory Coupling at Baseline: The Pulse-Breath Dialogue

At rest, every inhale should:
- Increase HR
- Decrease vagal activity

Every exhale should:
- Decrease HR
- Increase vagal activity

Loss of this coupling is one of the earliest signs of parasympathetic decline.

Baseline Respiratory-Coupling Patterns

Pattern	Mechanism	Meaning
Smooth oscillation	Intact RSA*	Healthy vagal function
Flattened pattern	Weak RSA	Early parasympathetic injury
Irregular swaps	Timing mismatch	Central dysautonomia
Sympathetic contamination	Reduced oscillation	Stress-related noise or compensation

Respiratory Sinus Arrhythmia (RSA) is the normal, healthy pattern where your heart rate speeds up when you inhale and slows down when you exhale.

Breath-to-beat coupling is one of the first autonomic functions to erode—and one of the most reliable early markers of disease.

Section 7—HRV at Baseline: The Reserve Gauge

High HRV at rest means:
- Strong vagal tone
- Flexible response mechanisms
- Adequate reserve

Low HRV means:
- Fatigue
- Poor recovery
- Autonomic exhaustion
- Vulnerability to upright stress
- Susceptibility to heat and exercise crashes

Resting HRV Interpretation

Finding	Meaning	Clinical Significance
High RMSSD	Strong parasympathetic tone	Excellent reserve
Low RMSSD	Vagal impairment	Predictably poor orthostatic performance
Low SDNN	Global reduction	Multisystem involvement
High LF/HF	Sympathetic dominance	Compensation or stress

HRV is not a lifestyle metric. It is the autonomic system's credit report.

Section 8—Resting State as a Predictor of Upright Performance

When the resting state is abnormal, the upright state will be worse—guaranteed.

Resting → Upright Predictive Patterns

Resting Finding	Upright Consequence	Mechanism
Low HRV	Exaggerated HR rise	Weak parasympathetic braking
High resting HR	Early POTS physiology	Compensatory baseline
BP instability	Orthostatic hypotension	Baroreflex weakness
Sympathetic noise	Hyperadrenergic response	Overdrive on standing
Low pulse pressure	Presyncope	Poor stroke volume

The resting state is the teaser trailer for the full autonomic movie. If the preview is chaotic, the feature film will be worse.

Science Snapshot: The Resting State Map

Rest → Reflex → Mechanism

Baseline Finding	Reflex Test Result	Mechanism Revealed
Low HRV	Weak deep breathing	Vagal impairment
Resting tachycardia	Excessive HR rise	Low stroke volume
Wide pulse pressure	Exaggerated heat response	Vascular underactivation
Labile BP	Abnormal Valsalva	Baroreflex disorder
Fixed HR	Blunted respiratory coupling	Autonomic rigidity

Most diagnoses become obvious here—long before tilt tables or blood draws.

Translation—What This Chapter Really Means
- Resting physiology is the ANS's default setting—its "true self."
- If the resting state is abnormal, the upright state will be catastrophic.

- High resting HR does not mean anxiety. It means compensation.
- Low resting HR can be athletic or pathologic—the HRV tells you which.
- Instability at rest foreshadows collapse during stress, heat, posture, or exertion.
- Baseline doesn't lie. Everything else is just elaboration.

Tilt Tip—How to Use This Knowledge Right Now

- Track resting HR + HRV daily—they reveal more than symptoms do.
- Watch for morning abnormalities—morning baseline predicts daily stability.
- If your resting state is unstable, plan upright activities strategically.
- Use resting measures to detect flares early—before major crashes occur.
- Bring baseline trends to clinicians—resting instability is diagnostic, not dramatic.

CHAPTER 6
THE DEEP BREATHING TEST: THE PARASYMPATHETIC PLAYBOOK

Where the vagus nerve stops pretending it's fine and finally shows you what it's really capable of—or not.

Introduction—The Test Everyone Thinks They Understand but Almost No One Actually Does

Deep breathing in autonomic testing is the physiologic equivalent of asking the parasympathetic nervous system, "So... do you actually do anything around here?"

This test exposes vagal truth with the subtlety of a subpoena. No drama. No theatrics. No interpretation bias.

Either the vagus nerve modulates heart rate gracefully, coordinates respiratory–cardiac coupling, updates RR intervals like a pro, and produces a smooth, sinusoidal dance—or it delivers a jagged, exhausted curve that screams, "I have absolutely nothing left to give!"

This chapter explains why deep breathing is not a meditation exercise, not a wellness suggestion, and not optional. It is the diagnostic backbone of parasympathetic function—and the earliest sign that the autonomic system can no longer improvise.

Section 1—What Deep Breathing Actually Measures (Hint: Not Calmness)

Deep breathing measures respiratory sinus arrhythmia (RSA)—the heart's ability to adjust its rhythm in response to each inhalation and exhalation.

This test does NOT measure relaxation, "mind-body harmony," anxiety control, spiritual alignment, or breathing technique.

This test measures vagal integrity, baroreflex coordination, beat-to-beat adaptability, and parasympathetic strength.

RSA is the vagus nerve's signature. When it's strong, the system is adaptable. When it's weak, the system is fragile.

Healthy Deep Breathing: What It Should Look Like

Breathing Phase	Heart Rate Response	What It Means
Inhale	HR rises	Vagal withdrawal works
Exhale	HR falls	Vagal activation works
Peak–trough difference	≥15 bpm	Robust parasympathetic modulation
Pattern	smooth wave	Timing intact, coordination excellent

A healthy vagus nerve produces a curve so smooth it could be mistaken for a graphic design demo.

Dysfunctional Deep Breathing: What It Looks Like Instead

Pattern	Interpretation	Mechanism
Flatline	No modulation	Vagal failure
Shallow waveform	Weak modulation	Early parasympathetic decline
Irregular timing	Central dysfunction	Impaired reflex coordination
Excessive HR rise	Sympathetic contamination	Parasympathetic withdrawal + overdrive

The deep breathing test is the parasympathetic polygraph. It tells the truth even when everything else looks normal.

Section 2—Why Parasympathetic Function Matters More Than People Realize

Parasympathetic tone is not a luxury. It is the foundation of digestion, recovery, immune stability, restorative sleep, emotional regulation, blood pressure stability, heart rate modulation, and inflammation control.

When vagal function declines, the entire body becomes less coordinated and more reactive.

Parasympathetic Failure → Predictable Consequences

Domain	Failure Outcome	Real-Life Impact
Cardiac	Poor HR braking	Tachycardia, excessive HR spikes
Vascular	Weak tone modulation	Labile BP, blood pooling
GI	Dysmotility	Nausea, bloating, slow transit

Domain	Failure Outcome	Real-Life Impact
Sleep	Non-restorative	Waking unrefreshed, morning crashes
Immune	Exaggerated inflammatory response	Flares after minor stress
Cognition	Reduced perfusion modulation	Brain fog, fatigue

Vagal decline is not subtle. It is the beginning of system-wide instability.

Section 3—Anatomy of the Deep Breathing Curve: The Autonomic Signature

This test captures three major features:

1. Amplitude (How much does HR change?)
High amplitude:
- Strong vagal activity
- Healthy cardiac modulation

Low amplitude:
- Impaired parasympathetic tone
- Early dysfunction

Zero amplitude:
- Severe vagal failure

2. Timing (Does the change happen when it should?)
Healthy timing:
- HR rises exactly on inhalation
- HR falls exactly on exhalation

Abnormal timing indicates:
- Brainstem dysfunction
- Baroreflex delay
- Central autonomic instability

3. Pattern (Is the curve smooth or chaotic?)

Smooth pattern:
- Well-coordinated system
- High adaptability

Chaotic pattern:
- Sympathetic contamination
- Stress physiology
- Compensation attempts

Deep Breathing Interpretation Table

Finding	Likely Mechanism	Diagnostic Clue
Low amplitude	Vagal impairment	Common in POTS, post-viral states
Delayed trough	Baroreflex delay	Early central dysfunction
Irregular oscillation	Timing failure	Mixed autonomic dysfunction
No oscillation	Severe failure	Advanced neuropathy or central disease
Overshooting HR rise	Sympathetic overdrive	Hyperadrenergic physiology

One test. One minute. A thousand insights.

Section 4—The Most Misinterpreted Curve in Autonomic Medicine

Deep breathing curves are constantly misread because clinicians confuse breathing effort with physiologic response.

Here's the truth: you can take perfect deep breaths and still show vagal failure, and you can take sloppy breaths and still generate high-quality variability. Technique does not override parasympathetic impairment.

The vagus nerve isn't concerned with your breathwork form.

Common Misinterpretations (and Why They're Wrong)

Misinterpretation	Why It's Incorrect
"You didn't breathe deeply enough."	Breathing technique does not create variability.
"You were too stressed during the test."	Stress does not eliminate vagal oscillation—it reduces amplitude slightly.
"Try again, but relax more."	Parasympathetic impairment does not improve with relaxation.
"Breathing exercises will fix this."	Exercises strengthen respiratory muscles, not vagal tone.

The deep breathing test is immune to excuses.

Section 5—How Deep Breathing Predicts Upright Tolerance

If the parasympathetic system can't modulate heart rate while you're sitting still, it will absolutely fail under upright stress.

Predictive Power of Deep Breathing

Deep Breathing Finding	Upright Consequence
Low amplitude	Exaggerated HR rise, unstable BP
Delayed pattern	Delayed compensation standing
No oscillation	Severe orthostatic intolerance
Irregular curve	Unpredictable symptoms

Vagal strength is the determinant of stability after meals, recovery from heat, HR control under stress, exercise tolerance, morning recovery, and resilience to daily stimuli.

If vagal modulation is weak, everything becomes harder.

Section 6—Deep Breathing vs. HRV: The Cousins That Speak the Same Truth

Deep breathing is the structured, forced version of HRV. They reveal similar failures in different ways.

HRV vs. Deep Breathing

Domain	HRV Measures	Deep Breathing Measures
Tone	Ongoing vagal reserve	Forced vagal activation
Adaptability	Moment to moment	Inhale/Exhale coupling
Rhythm	Global pattern	Specific reflex
Vulnerability	Baseline	Extreme challenge

HRV says, "Here's your daily parasympathetic capacity." Deep breathing says, "Show me what you can do when I push you."

Both matter. Both diagnose differently. Both must be read together.

Section 7—Medication Effects: The Parasympathetic Impersonators

Some medications pretend to enhance vagal tone—or suppress it.

Medications That Lower Amplitude (False Vagal Weakness)

Class	Effect
Anticholinergics	Suppress vagal activation
Stimulants	Increase sympathetic noise
TCAs	Suppress parasympathetic tone

Medications That Increase Amplitude (Artificial Vagal Support)

Class	Effect
Benzodiazepines	Cosmetic amplitude improvement
Beta-blockers	May exaggerate parasympathetic appearance

This is why medication history is not optional. It is part of the interpretation.

Science Snapshot—What Deep Breathing Actually Diagnoses

Three Metrics → Three Mechanisms

Marker	Meaning	Mechanism
Low amplitude	Poor vagal tone	Parasympathetic impairment

Marker	Meaning	Mechanism
Delayed trough	Poor baroreflex timing	Central dysfunction
Irregular curve	Sympathetic overactivity	Contamination or compensation

These three alone identify 70% of parasympathetic disorders.

Translation—What This Chapter Really Means

- Deep breathing is the single best test of vagal strength.
- Your ability to exhale and reduce heart rate is more important than your resting HR.
- If the curve is flat, the parasympathetic system is not working.
- If the curve is delayed, the baroreflex is struggling.
- If the curve is irregular, the sympathetic system is interfering.
- This test doesn't measure breathing skill—it measures autonomic truth.

Tilt Tip—How to Use Deep Breathing in Real Life

- If deep breathing amplitude is low, expect exaggerated HR responses under stress.
- If timing is delayed, standing will feel uneasy, with delayed dizziness.
- If the curve is irregular, stress tolerance will be unpredictable.
- Track how meals, sleep, hydration, heat, and stress affect breathing-induced HR changes.
- Use deep breathing test results to explain symptoms to clinicians—vagal weakness is measurable, not emotional.

CHAPTER 7

VALSALVA MANEUVER: A SYMPATHETIC STRESS TEST WITHOUT A TREADMILL

Because nothing reveals autonomic failure faster than asking the body to briefly implode and recover gracefully.

Introduction—The Test That Turns Your Chest Into a Pressure Cooker for Science

The Valsalva maneuver is the autonomic equivalent of a high-stakes interrogation.

It is controlled cardiovascular chaos—a brief internal crisis, a pressure-induced blood-flow disaster, a stress test without the treadmill, and the fastest way to expose sympathetic and baroreflex dysfunction.

During Valsalva, you strain for 15–20 seconds—bearing down like you're trying to blow air through a brick wall—which sends intrathoracic pressure sharply upward. It's the same basic maneuver fighter pilots use to stay conscious at 9 Gs—except you're doing it just to stay upright in a retail superstore.

This drastically reduces venous return, challenges the cardiovascular system, and forces the autonomic nervous system to either respond with spectacular competence or confess catastrophic weakness.

This is the only test where physiology stops lying.

Section 1—What the Valsalva Maneuver Really Tests

The Valsalva maneuver tests four key domains simultaneously:

1. Sympathetic activation strength
2. Baroreflex timing and coordination
3. Vascular responsiveness
4. Parasympathetic braking ability (recovery)

Most autonomic tests look at one branch. Valsalva interrogates all of them. It is a four-phase challenge that forces the autonomic system to react, compensate, stabilize, and recover. If any part of this sequence falters, the entire control loop falls apart.

Section 2—The Four Phases: The Autonomic Playbook Exposed

The Valsalva maneuver is divided into four phases, each revealing a different aspect of autonomic function. This is not optional scientific trivia. It is the key to understanding the mechanism behind every dysautonomia pattern.

Phase I—Mechanical Pressure Spike

The only phase that requires zero autonomic skill.

- You begin straining.
- Pressure in the chest spikes.
- Blood pressure rises abruptly due to physics, not nerves.

If Phase I is abnormal, it's usually equipment error or positioning—not physiology.

Phase II—The Sympathetic Exam: Compensation Under Fire
This is where the autonomic system must fight back.

What Should Happen:
As intrathoracic pressure blocks venous return:
- Stroke volume drops
- Blood pressure falls
- The sympathetic nervous system should immediately respond
- HR rises
- Peripheral vasoconstriction increases
- BP stabilizes

Phase II is the sympathetic system's moment of truth.

What Failure Looks Like:
- HR barely rises → sympathetic underactivation
- HR skyrockets → sympathetic desperation
- BP continues falling → vascular incompetence
- BP oscillates → baroreflex miscoordination

This phase alone can diagnose half of all autonomic disorders.

Phase III—Release Drop
The system reveals timing under pressure.

As the strain ends:
- Intrathoracic pressure drops
- Venous return briefly dips
- BP falls

Phase III is where delayed sympathetic compensation becomes obvious. A healthy system anticipates this dip. A dysfunctional system reacts late—or not at all.

Phase IV—Overshoot and Brake: The Parasympathetic Finale
If the sympathetic system did its job in Phase II, Phase IV shows it.

Healthy Phase IV:
- BP overshoots baseline
- Baroreflex activates
- Parasympathetic braking kicks in
- HR drops sharply

Beautiful, crisp, tightly timed physiology.

Abnormal Phase IV Patterns:
- No overshoot → sympathetic failure
- Weak overshoot → vascular insufficiency
- No HR drop → vagal impairment
- Delayed HR drop → baroreflex timing issue

Phase IV is the single most sensitive marker of vagal function after deep breathing.

Section 3—Interpreting Valsalva: The Patterns That Matter

Pattern 1—Blunted Phase II HR Rise
- Mechanism: sympathetic underactivation
- Meaning: early neuropathy, ganglionic failure

Pattern 2—Massive HR Spike With Weak BP Compensation
- Mechanism: hyperadrenergic physiology
- Meaning: poor vascular response, excessive norepinephrine output

Pattern 3—No Phase IV Overshoot
- Mechanism: baroreflex failure
- Meaning: central autonomic disorder or advanced neuropathy

Pattern 4—Oscillating BP Waves
- Mechanism: delayed baroreflex feedback
- Meaning: coordination breakdown, central dysautonomia

Pattern 5—Delayed Recovery After Phase IV
- Mechanism: parasympathetic exhaustion
- Meaning: poor vagal braking capacity

Valsalva patterns are not vague—they are fingerprints.

Section 4—Valsalva and Orthostatic Physiology: The Missing Link

Valsalva replicates the exact physiology of standing, just without gravity.
- Standing: Venous return falls → BP drops → SNS activates → HR rises → BP stabilizes
- Valsalva: Venous return falls → BP drops → SNS activates → HR rises → BP stabilizes

If Valsalva is abnormal, upright physiology will be worse.

Valsalva → Upright Prediction Table

Valsalva Abnormality	Upright Outcome
Weak Phase II	Orthostatic hypotension
Massive HR spike	POTS physiology
No Phase IV	Baroreflex failure
Oscillations	Upright instability
Slow recovery	Prolonged tachycardia

Valsalva is the orthostatic dress rehearsal. Standing up is the main performance.

Section 5—Valsalva Ratio: The Most Misused Metric in Autonomic Medicine

The Valsalva Ratio (VR) compares peak HR during the strain to trough HR during recovery. In theory, VR > 1.2 → normal. In practice, VR oversimplifies everything.

Why VR Alone Is Useless

Problem	Why It Matters
Ignores BP curves	Can miss baroreflex failure
Misreads tachycardia	High HR spikes inflate the ratio
Overemphasizes HR	Ignores vascular response
Doesn't evaluate timing	Hides coordination issues

Clinicians who rely solely on VR are missing 70% of the pathology. Reading the Valsalva correctly means evaluating blood pressure morphology, heart rate timing, overshoot presence, amplitude, slope, and coordination—not just a single number.

Section 6—Hyperadrenergic States Exposed by Valsalva

One of Valsalva's greatest strengths is exposing hyperadrenergic physiology without requiring the patient to stand.

Signs of Hyperadrenergic Behavior in Valsalva

Finding	Mechanism	Meaning
Huge Phase II HR spike	SNS overdrive	Compensatory escalation
Poor BP stabilization	Vascular incompetence	Poor peripheral resistance
Excessive Phase IV overshoot	Exaggerated baroreflex	Hyperactive response
Tremulous HR pattern	Noise contamination	Adrenaline-driven instability

Hyperadrenergic syndromes often masquerade as anxiety. Valsalva removes the disguise.

Section 7—Neuropathic Patterns in Valsalva

Neuropathic autonomic disorders produce their own Valsalva signature.

Signs of Neuropathy in Valsalva

Finding	Mechanism
Little to no HR rise	Sympathetic denervation
Low or absent BP stabilization	Loss of vascular innervation
No Phase IV overshoot	Baroreflex failure
Flat recovery	Parasympathetic involvement

This pattern is profoundly different from hyperadrenergic states, even if the symptoms overlap.

Section 8—The Power of the Phase Map: A Diagnostic Summary

The Four-Phase Interpretation Table

Phase	Expected	Abnormality	Meaning
I	Mechanical spike	Absent	Instrumentation or positioning
II	HR rise + BP stabilization	Weak/Loss	Sympathetic failure
III	Brief dip	Absent/Delayed	Coordination issue
IV	Overshoot + HR brake	Absent	Baroreflex or vagal failure

Four phases. Four mechanisms. Immediate clarity.

Science Snapshot—What the Valsalva Actually Proves

Valsalva Diagnostic Hierarchy

Failure	Mechanism	Interpretation
No compensation	Sympathetic underactivation	Neuropathic physiology
Excessive HR rise	Sympathetic overdrive	Hyperadrenergic physiology
No overshoot	Baroreflex failure	Central vs. Neuropathic
No HR braking	Vagal failure	Parasympathetic impairment
Delay in transition	Timing disorder	Brainstem or baroreflex

Valsalva doesn't just evaluate the ANS—it profiles it.

Translation—What This Chapter Really Means

- Valsalva recreates the worst parts of standing without requiring you to stand.
- Phase II exposes sympathetic strength or desperation.
- Phase IV exposes parasympathetic braking capacity.
- BP curves matter more than HR curves.
- Hyperadrenergic responses look nothing like neuropathic responses—even if symptoms are identical.
- Valsalva is the only test where the entire ANS is forced to reveal itself honestly.

Tilt Tip—How to Use Valsalva Insights in Real Life

- If Phase II is weak, expect blood pressure drops during upright activities.
- If Phase IV is absent, push for baroreflex evaluation.
- If HR skyrockets but BP doesn't stabilize, you are compensating—not overreacting.
- If recovery is slow, anticipate long post-exertional crashes.
- If Valsalva is abnormal, upright life will be chaotic—build pacing strategies accordingly.

CHAPTER 8
STAND OR TILT: ORTHOSTATIC WARFARE

Because nothing exposes autonomic dysfunction faster or more brutally than asking a human being to defy gravity.

Introduction—The Battle You Fight Every Time You Stand Up

Standing upright is the most underrated physiologic challenge in human existence.

It is the daily stress test we perform without thinking—until the autonomic system fails, and suddenly standing becomes a negotiation, a gamble, a tactical engagement, and occasionally a slow-motion collapse broadcast in high definition.

Orthostatic testing—whether standing or using the tilt table—is not just "checking vitals while upright." It is a controlled battle simulation.

This chapter explains why upright posture transforms dysautonomia from a "mystery" into a mathematically predictable collapse pattern—and why any clinician who ignores orthostatic physiology is diagnosing in the dark.

Section 1—Why Standing Is Such a Physiologic Nightmare

When a human being stands upright, gravity immediately pulls 500–1000 mL of blood toward the legs and abdomen. That is one-fifth of your entire blood volume suddenly abandoning your brain.

A healthy autonomic nervous system responds instantly by constricting blood vessels, increasing heart rate, boosting cardiac output, raising peripheral resistance, maintaining cerebral perfusion, and holding everything together seamlessly.

This entire process takes 5 seconds in healthy physiology. In dysautonomia, it takes too long, too much effort, or it doesn't happen at all.

Normal Orthostatic Response

Parameter	Normal Change	Why
HR	+10–20 bpm	Compensate for reduced venous return
SBP (systolic blood pressure)	Stable (±5)	Adequate vascular tone
DBP (diastolic blood pressure)	Slight increase	Peripheral constriction
Pulse pressure	Stable	Preserved stroke volume
Symptoms	None	Brain adequately perfused

Orthostatic Intolerance Response

Parameter	Dysfunctional Change	Meaning
HR	+30–70 bpm	Compensating for vascular failure
BP	May drop, spike, or oscillate	Baroreflex instability
Pulse pressure	Narrows	Falling stroke volume
Symptoms	Dizziness, heat, tremor, fog	Poor cerebral perfusion

This is not emotional sensitivity or "overreacting to standing." It is hemodynamic physics under duress.

Section 2—Standing vs. Tilt Testing: What Each Actually Tells You

Standing reveals how your physiology behaves in the real world. Tilt testing removes the muscle pump so the autonomic system can't perform any quiet "extra credit." They answer different questions—but you don't need both to reach a meaningful conclusion.

1. Standing Test: Real-World Performance
Strengths:
- Captures venous pooling the moment gravity takes charge
- Highlights exercise intolerance without dramatics
- Includes the inevitable muscle-pump interference of daily life
- Reflects symptoms as they appear outside the clinic

Limitations:
- Patients shift, brace, and micro-adjust even when they swear they're standing still
- Muscle contractions offer partial compensation
- The environment is less controlled, more "clinical reality" than "clinical ideal"

2. Tilt Test: Pure Autonomic Truth
Strengths:
- Eliminates the muscle pump entirely
- Prevents compensatory shifting
- Allows precise, reproducible angle control
- Produces cleaner HR/BP patterns
- Makes phase transitions easier to identify

Limitations:
- Exists in a mildly artificial testing environment
- May provoke stronger symptoms in severe cases

Standing shows real-world function. Tilt shows uncompromised physiology. Choose the test that fits the question—not because both are required, but because each offers a different piece of the story.

Section 3—The Phases of Orthostatic Testing

Upright physiology breaks into three predictable phases:

Phase 1—Initial Orthostatic Response (0–30 seconds)
The fastest reflexes activate:
- Sympathetic surge
- Vascular constriction
- Heart rate increase

Symptoms here indicate early autonomic failure.

Phase 2—Intermediate Phase (30 seconds–3 minutes)
Baroreflex takes over:
- HR stabilizes
- BP normalizes
- Cerebral perfusion restored

Problems in this phase indicate baroreflex dysfunction.

Phase 3—Steady-State Phase (3–10 minutes)
The system either adapts and stabilizes or continues to deteriorate, revealing patterns consistent with postural orthostatic tachycardia syndrome (POTS), orthostatic intolerance (OI), orthostatic hypotension (OH), or syncope.

Section 4—Orthostatic Phenotypes: The Four Big Ones

1. Postural Orthostatic Tachycardia Syndrome (POTS)
Key Features:
- HR increases ≥30 bpm within 10 minutes
- No significant drop in BP
- Stroke volume declines
- Sympathetic compensation escalates
- Brain perfusion becomes precarious

Why it happens: Vascular underactivation + venous pooling → low preload → compensatory tachycardia.

How it feels:
- Racing heart
- Tremors
- Heat rising
- Sweating
- Dizziness
- Cognitive fog
- "I need to sit NOW"

Not anxiety. Not deconditioning. Not sensitivity. Compensation in action.

2. Orthostatic Hypotension (OH)
Key Features:
- SBP drop ≥20 mmHg or DBP ≥10 mmHg
- Weak sympathetic activation
- Inadequate vasoconstriction
- Cerebral perfusion collapse

Mechanism: Sympathetic failure → vasculature refuses to tighten → BP crashes.

Symptoms:
- Gray-out
- Tunnel vision
- Weakness
- Near-syncope
- Actual syncope

This is not subtle. This is vascular rebellion.

3. Delayed Orthostatic Hypotension
Key Features:
- BP normal at first
- Drops after prolonged standing
- Often missed on short tests

Mechanism: Slow sympathetic decompensation + venous pooling accumulation.

Symptoms:
- Worsening over minutes
- Fatigue
- Mental fog
- Instability

This is the stealth version of OH.

4. Neurocardiogenic Syncope (NCS)/Vasovagal Syncope
Key Features:
- Prolonged compensation
- Sudden vagal surge
- HR drops
- BP collapses
- Patient drops

Mechanism: The autonomic system misreads prolonged standing as a threat → overcorrects → shuts down.

Symptoms:
- Nausea
- Sweating
- Visual fade
- Sudden collapse

This is not fainting from fear. This is your body hitting CTRL+ALT+DELETE.

Section 5—The Orthostatic Variables That Clinicians Ignore (But Shouldn't)

1. Pulse Pressure (PP)
One of the single most revealing metrics in orthostatic testing.
- PP = SBP (systolic blood pressure) − DBP (diastolic blood pressure)

Interpretation

PP Pattern	Meaning
Stable	Adequate stroke volume
Narrowing	Falling stroke volume (POTS hallmark)
Widening	Peripheral vasodilation

If PP narrows while HR rises, that's not anxiety—it's compensation.

2. Cerebral Perfusion Symptoms

Even when blood pressure is normal, patients may experience brain fog, dizziness, disorientation, fatigue, or speech difficulty because cerebral blood flow depends on pulse pressure, autonomic timing, and venous return—not BP alone.

3. Heart Rate Morphology

Not just how much HR increases, but how the curve behaves:

Morphology	Meaning
Smooth rise	Healthy compensation
Jagged spikes	Sympathetic instability
Biphasic rise	Delayed baroreflex
Plateau	Chronic compensation
Escalating climb	Inadequate stabilization

The shape is diagnostic.

Section 6—The Test Environment Matters More Than Anyone Admits

Orthostatic testing can be influenced by room temperature, emotional stress, hydration, medications, recent meals, and time of day.

Morning orthostatic testing reveals the worst pooling, lowest autonomic reserve, highest tachycardia potential, and most reliable patterns. Afternoon testing often understates severity. Heat exposure aggravates vasodilation, pooling, and HR rise.

Testing should control these variables—but rarely does.

Section 7—Pooling: The Hidden Enemy

Blood pooling isn't subtle. It's measurable. It's visible. It's diagnosable.

Signs of Lower-Body Pooling

Sign	Mechanism
Leg heaviness	Decreased venous return
Purple mottling	Venous stasis
Cold extremities	Vasoconstriction failure
Lower body pressure	Blood accumulation
Orthostatic worsening	Insufficient compensation

Pooling is the backbone of POTS physiology—and the foundation of orthostatic intolerance.

Section 8—Breathing Artifacts: The Reason HR Isn't Everything

A patient breathing rapidly under stress may produce HR spikes, irregular intervals, and false tachycardia patterns.

But BP curves tell the truth. If blood pressure is stable, pulse pressure is intact, and there are no cerebral symptoms, then heart rate behavior reflects noise rather than collapse.

In dysautonomia, both HR and BP are meaningful. In anxiety, only HR is.

Section 9—The Patterns That Reveal Mechanism at a Glance

Orthostatic Interpretation Table

Pattern	Mechanism	Interpretation
HR↑, BP stable	POTS	Compensation for low SV (stroke volume)
HR↑, PP ↓	Low stroke volume	Pooling or dysautonomia
BP ↓ early	Sympathetic failure	OH
BP ↓ late	Delayed OH	Baroreflex fatigue
HR↓ + BP↓	Vasovagal	Vagal surge
BP oscillation	Baroreflex instability	Central dysfunction
HR↑ + tremor	Hyperadrenergic	Excessive NE surge

This table is the orthostatic Rosetta Stone.

Science Snapshot—Gravity vs. Autonomic Function

Standing → What Should Happen → What Goes Wrong in Dysautonomia

Event	Healthy Response	Dysfunctional Response
Blood shifts downward	Vasoconstriction	Pooling
Venous return drops	HR↑ modestly	HR↑ dramatically
Cerebral perfusion threatened	BP maintained	BP drops or oscillates
Oxygen delivery stabilizes	No symptoms	Dizziness, fog, tremor
Steady state reached	Asymptomatic	Exhaustion, collapse

Standing is a negotiation with gravity. Dysautonomia shows up to the negotiation table with no leverage and even less dignity.

Translation—What This Chapter Really Means

- Standing is a physiologic battlefield—not a simple posture.
- Tilt testing exposes the system's ability to compensate without muscular help.
- Pulse pressure is one of the most important metrics in autonomic testing.
- HR alone does not define orthostatic intolerance—BP and PP matter more.
- POTS is not "anxiety." Orthostatic hypotension is not "dehydration."
- All orthostatic disorders follow predictable patterns that correlate with mechanism.
- If upright physiology is unstable, the patient's symptoms are not vague—they are measurable.

Tilt Tip—How to Apply Orthostatic Knowledge Today

- Track symptoms by phase: early, mid, and late standing.
- Use pulse pressure trends to understand crash triggers.
- Perform morning orthostatic checks—they reveal the truth.
- Recognize pooling signs early and intervene before collapse.
- Use standing + tilt results to differentiate POTS from OH and vasovagal patterns.
- When clinicians dismiss symptoms, presenting BP + PP curves changes the entire conversation.

CHAPTER 9
SUDOMOTOR ELECTROCHEMISTRY: SWEAT AS A DIAGNOSTIC TOOL

Because your sweat glands know more about your nerves than most clinicians do.

Introduction—The Secret Autonomic System You Never Think About Until It Fails

Of all the autonomic subsystems, the sudomotor pathway is the quiet background intern—doing thankless work, only noticed when something catastrophic happens.

When healthy, it regulates thermoregulation, electrolyte balance, surface cooling, peripheral nerve signaling, and heat stress tolerance.

When unhealthy, it produces heat intolerance, temperature dysregulation, flushing, cold limbs, burning skin, and sweat that acts unpredictably—sometimes like a broken sprinkler system.

Sudomotor testing is the cleanest way to detect small fiber neuropathy, early dysautonomia, and subtle autonomic decline long before other tests catch abnormalities.

This chapter teaches you how to interpret sweat like the diagnostic confession it truly is.

Section 1—What Sudomotor Function Actually Measures

Sudomotor testing assesses postganglionic sympathetic cholinergic fibers—the smallest, most fragile, and earliest-to-fail autonomic fibers in the peripheral nervous system.

These fibers activate sweat glands, mediate vasodilation, regulate thermal balance, interact with nociceptive pathways, and are the first to show damage in early neuropathies.

When these fibers fail, the symptoms are not vague—they follow a predictable physiologic pattern.

Sudomotor Function Explains Common Dysautonomia Complaints

Symptom	Mechanism
Heat intolerance	Reduced evaporative cooling
Flushing	Unstable vasodilation
Cold hands/feet	Vasoconstriction imbalance
Tingling/Burning	Small fiber involvement
Unexplained fatigue in warm rooms	Impaired cooling efficiency

If you can't sweat properly, you can't regulate temperature properly. And if you can't regulate temperature properly, your autonomic reserves deplete instantly.

Section 2—Sudomotor Anatomy: The Underappreciated Network

Sweat glands are activated by sympathetic cholinergic fibers (yes—sympathetic nerves using acetylcholine, not norepinephrine).

Sudomotor Pathway
1. CNS initiates thermoregulatory command
2. Preganglionic sympathetic neurons transmit signal
3. Synapses occur in the sympathetic chain
4. Postganglionic cholinergic fibers innervate sweat glands
5. Sweat glands respond with appropriate fluid release

This is a postganglionic test, meaning it identifies the location of dysfunction with surprising precision.

Section 3—What Sudomotor Electrochemistry Actually Measures

Modern sudomotor testing (e.g., electrochemical skin conductance) evaluates chloride movement, sweat gland ion transport, skin conductance, peripheral nerve integrity, thermoregulatory capacity, and localized small fiber function.

It does not measure hydration, stress-induced sweat, emotional sweating, deodorant effectiveness, or fitness level.

This test probes nerves, not lifestyle.

Measurable Sudomotor Parameters

Parameter	Meaning
Skin conductance	Sweat gland output efficiency
Asymmetry	Localized small fiber neuropathy
Low global conductance	Generalized autonomic impairment
Distal > proximal weakness	Length-dependent neuropathy
Regional deficits	Focal nerve injury

Sudomotor dysfunction is often the earliest sign of autonomic decline.

Section 4—The Patterns of Sudomotor Dysfunction

Sudomotor abnormalities fall into predictable patterns that directly correlate with mechanism.

Pattern 1—Length-Dependent Neuropathy
The classic "worse in the feet" pattern.
- Distal sweat loss
- Proximal relative preservation
- Symmetric deficits

Mechanism: Progressive small fiber neuropathy (diabetes, chemotherapy, autoimmune diseases).

Pattern 2—Non–Length-Dependent Patchy Loss
Randomized dysfunction across regions.

Mechanism:
- Immune-mediated dysautonomia
- Post-viral small fiber neuropathy
- Genetic autonomic disorders

Patchiness matters. It suggests immune dysregulation, not degenerative patterns.

Pattern 3—Global Sudomotor Failure
No sweat. Nowhere.

Mechanism:
- Severe autonomic neuropathy
- Ganglionic dysfunction
- Advanced autoimmune autonomic failure

This is a major diagnostic marker of advanced disease.

Pattern 4—Hyperactive Sudomotor Output
Excessive sweating in warm or even cool conditions.

Mechanism:
- Sympathetic overdrive
- Compensatory response to proximal failure
- Heat-triggered exaggerated reflex

Not anxiety. Not personality. Not "overreacting." Compensation.

Section 5—Sudomotor Testing vs. Skin Biopsy vs. QSART

Clinicians often confuse these tests or assume they are interchangeable. They are not.

Comparison Table

Test	Measures	Strengths	Limitations
Sudomotor electrochemistry	Chloride flux + conductance	Quick, painless, postganglionic specificity	Semi-quantitative
QSART	Gland activation via acetylcholine	Highly sensitive	Time-intensive
Skin biopsy	Small fiber density	Gold standard for structure	Does not measure function
Thermoregulatory sweat test	Whole-body mapping	Precise patterning	Messy & heat-triggering

Sudomotor electrochemistry is the fastest, cleanest window into autonomic small fiber integrity.

Section 6—Sudomotor Failure Predicts Heat Intolerance and Orthostatic Instability

Sweating is your primary cooling method. Without functional sweat glands, heat accumulates, blood vessels dilate, and cardiac output collapses.

Consequences of Sudomotor Dysfunction

Dysfunction	Impact	Outcome
Poor sweating	Poor heat dissipation	Overheating, fatigue
Excess vasodilation	Reduced venous return	Orthostatic intolerance
Small fiber loss	Impaired vascular tone	Dizziness, instability
Altered ion transport	Neurologic hypersensitivity	Burning, tingling sensations

Heat intolerance is not subjective—it is quantifiable through sudomotor dysfunction.

Section 7—Sudomotor Asymmetry: Why Uneven Sweating Matters

Asymmetric deficits indicate localized nerve injury, early autoimmune disease, segmental small fiber neuropathy, or early axonal damage.

This is diagnostic gold. Asymmetry helps identify where the autonomic system falters first.

Asymmetry Interpretation

Asymmetric Pattern	Meaning
Unilateral deficit	Localized neuropathy
Distal > Proximal	Length-dependent process

Asymmetric Pattern	Meaning
Proximal > Distal	Ganglionic involvement
Diagonal pattern	Immune or genetic failure

The sweat gland map is a nerve-function blueprint.

Section 8—Sudomotor Patterns in Dysautonomia: The Mechanistic Link

Sudomotor testing is not just about sweat. It correlates directly with HRV abnormalities, Valsalva phases, resting state instability, POTS physiology, orthostatic intolerance, and temperature sensitivity.

Mechanistic Correlation Table

Sudomotor Finding	Autonomic Mechanism	Clinical Correlation
Global low output	Sympathetic failure	OH, severe OI
Distal low output	Small fiber neuropathy	POTS + sensory symptoms
Patchy pattern	Autoimmune	Post-viral dysautonomia
Hyper-sweating	Sympathetic overdrive	Hyperadrenergic POTS

If sweat is abnormal, upright physiology is rarely normal.

Section 9—Why Sudomotor Findings Are Often Dismissed (Incorrectly)

Common incorrect dismissals include: you were dehydrated; you were stressed; the room was too cool; it depends on how much water you drank; and sweating varies too much day to day.

These explanations are scientifically false. Sweating is not a mood. Sweating is not a preference. Sweating is not a lifestyle trait. Sweating is a nerve function.

Science Snapshot—The Sweat Gland Truth Table

Finding	Mechanism	Diagnosis
Global low conductance	Postganglionic failure	Severe autonomic neuropathy
Distal low conductance	Length-dependent small fiber loss	Diabetic/Idiopathic neuropathy
Patchy dysfunction	Immune/Ganglionic	Autoimmune dysautonomia
Hyper-conductance	Adrenergic excess	Hyperadrenergic states
Asymmetry	Focal nerve involvement	Segmental neuropathy

Sweat glands reveal nerve function long before symptoms do.

Translation—What This Chapter Really Means

- Sweat glands are autonomic sensors with extraordinary diagnostic value.
- Sudomotor testing identifies early nerve damage more reliably than HR or BP.
- Length-dependent patterns indicate neuropathy.
- Patchy patterns indicate immune dysfunction.
- Hyper-sweating indicates sympathetic compensation, not personality traits.
- Heat intolerance is measurable and not psychological.
- Sudomotor testing fills the diagnostic gap left by cardiovascular-only assessments.

Tilt Tip—How to Use Sudomotor Results in Real Life

- If sweating is low, avoid heat stress and build cooling strategies.
- If sweating is patchy, suspect autoimmune involvement and pursue targeted evaluation.
- If sweating is excessive, anticipate hyperadrenergic crashes.
- Combine sudomotor findings with HRV + Valsalva + standing tests to identify mechanisms.
- Bring sudomotor results to clinicians—objective data changes everything.

PART III
THE INTERPRETATION ENGINE: HOW TO READ THE BATTLEFIELD

Where data becomes clarity instead of chaos.

CHAPTER 10
NORMAL-NORMAL, NORMAL-ABNORMAL, ABNORMAL-ABNORMAL

Because test results don't lie, but people interpreting them sometimes convince themselves they do.

Introduction—The Three Combinations That Predict Everything

Autonomic testing produces a mountain of data, but if you zoom out far enough, every patient eventually falls into one of three patterns:

1. Normal-Normal
2. Normal-Abnormal
3. Abnormal-Abnormal

This deceptively simple framework is the backbone of interpretation. Once you understand what each pattern means—and more importantly, why it occurs—the entire autonomic landscape becomes clear.

Think of this system as the autonomic equivalent of medical triage:
- Who's fine.
- Who's compensating.
- Who's falling apart.

This chapter teaches you how to read test combinations the way an autonomic lab director does: ruthlessly, logically, and with zero tolerance for hand-waving.

Section 1—What "Normal" Actually Means in Autonomic Testing

"Normal" does not mean perfect. It does not mean symptom-free. It does not mean "nothing is wrong." Normal means the system is meeting physiologic demand without visibly failing.

A Deeper Definition of "Normal"

Category	Meaning
Physiologically normal	Responding within expected parameters
Functionally normal	Compensating well enough to maintain stability
Mechanistically normal	Pathways intact, no signal failure

Normal does not measure comfort. Normal measures performance.

In dysautonomia, many patients exhibit "normal" results because their system is overworking to compensate. That does not make it healthy; it makes it determined.

Section 2—What "Abnormal" Actually Means

Abnormal means the mechanism is failing, the compensation has collapsed, or the reflex is misfiring.

An abnormal result signals one of three possibilities: impaired signaling, impaired response, or impaired reserve.

Not all abnormalities are equal. Some reflect early dysfunction; others indicate system-wide collapse.

Types of Abnormality

Type	Meaning	Example
Amplitude abnormality	Weak response	Low HRV
Timing abnormality	Delayed reflex	Late Phase IV
Pattern abnormality	Inconsistent response	Oscillating BP
Direction abnormality	Reversed reflex	HR drop on standing
Morphology abnormality	Distorted curves	Ragged RSA

When interpreting autonomic testing, abnormalities matter—but combinations matter more.

Section 3—Pattern 1: Normal-Normal

Everything looks fine. Whether it actually is fine is another issue.

This pattern reflects a normal resting state, normal reflex testing, an orthostatic response within expected ranges, and intact sympathetic and parasympathetic function.

What Normal–Normal Really Means

Finding	Interpretation
All parameters normal	Mechanisms intact
Curves clean	No timing issues
BP + HR stable	Baroreflex functioning
HRV adequate	Vagal reserve intact

Patients in this category may experience symptoms from non-autonomic sources, symptoms arising outside the tested pathways, early dysautonomia that does not yet fail under laboratory conditions, or episodic instability not captured in a single snapshot.

The Mistake Clinicians Make

"This is normal, so nothing is wrong." Incorrect. Normal-Normal means the mechanism is intact during testing—nothing more.

When Normal–Normal Is Still Suspicious

Clinical Clue	Possible Mechanism
Symptoms only during heat	Thermoregulation issue not triggered
Symptoms only upright for long durations	Delayed orthostatic dysfunction
Symptoms only after meals	Splanchnic pooling
Symptoms only during exercise	Exertional autonomic failure
Symptoms only episodic	Intermittent baroreflex fatigue

Normal-Normal is reassuring but never definitive.

Section 4—Pattern 2: Normal-Abnormal

The most misdiagnosed category—and the most physiologically interesting.

This pattern is characterized by a normal resting state with one or more abnormal reflex tests or an abnormal standing test.

Normal-Abnormal indicates a system that looks fine at baseline but collapses under demand.

These patients pass the baseline tests with flying colors but fail when the ANS is pushed. This is the hallmark of early dysautonomia and many forms of POTS.

Mechanistic Interpretation

Component	State	Meaning
Baseline	Normal	Resting compensation adequate
Reflex response	Abnormal	Impaired parasympathetic or sympathetic reserve
Orthostatic response	Abnormal	Inadequate compensation under stress

Normal-Abnormal = compensation is working until you ask for performance. This is the "hidden dysfunction" group.

Why Normal-Abnormal Happens

1. Vagal impairment unmasked during deep breathing—baseline HRV may look okay, but amplitude collapses during forced breathing.

2. Sympathetic underactivation visible only during Valsalva Phase II—the system can hold resting tone but cannot generate force.

3. Baroreflex timing abnormalities—rest is stable, but compensation is delayed.

4. Orthostatic intolerance without baseline abnormalities—POTS, hyperadrenergic states, and early neuropathy all live here.

5. Early small fiber neuropathy—sudomotor abnormalities show up first while cardiac reflexes look intact.

Normal-Abnormal Clinical Portrait

These patients are functional in short bursts, unstable during prolonged upright time, affected by heat, affected by meals, affected by exercise, affected by stress, and chronically fatigued by overcompensation.

They look "normal" on paper and "fall apart" in real life—which is why so many clinicians mislabel them.

Normal-Abnormal Is the Most Common Dysautonomia Pattern. And the most frequently dismissed.

Section 5—Pattern 3: Abnormal–Abnormal

The system is not just struggling—it's failing regardless of circumstance.

This pattern shows an abnormal resting state, impaired reflexes, and abnormal orthostatic responses.

Nothing is compensating. Nothing is adapting. Nothing is stabilizing.

Abnormal-Abnormal Mechanistic Signature

Domain	Abnormality	Meaning
Baseline	HR, BP, HRV	Impaired resting autonomic tone
Deep breathing	Weak or absent RSA	Parasympathetic failure
Valsalva	Weak Phase II + poor overshoot	Sympathetic + baroreflex dysfunction
Standing	OH, severe OI, unstable PP	Advanced dysfunction

This is the clearest marker of significant autonomic disease.

Why Abnormal-Abnormal Happens
1. Advanced neuropathy—small fiber, autonomic ganglionic, or mixed.

2. Baroreflex failure—central pathway disruption or severe neuropathic damage.

3. Combined parasympathetic + sympathetic loss—both branches failing, resulting in instability at all levels.

4. Severe autoimmune autonomic ganglionopathy—patchy, widespread failure.

5. Chronic disease with autonomic decompensation—including metabolic, neurologic, or systemic inflammatory conditions.

Abnormal-Abnormal Clinical Portrait
These patients cannot maintain stability at rest, cannot mount reflex responses, cannot sustain upright posture, often have heat sensitivity, fatigue, pooling, dizziness, syncope, and show abnormalities across multiple testing domains.

Their autonomic system is not misfiring—it is malfunctioning.

Section 6—How to Use the Normal-Normal, Normal-Abnormal, Abnormal-Abnormal Framework
Interpretation Grid

Pattern	Meaning	Clinical Category
Normal-Normal	Intact mechanisms	Non-autonomic or early episodic
Normal-Abnormal	Functional until stressed	POTS, early OI, mild neuropathy
Abnormal-Abnormal	Global impairment	OH, severe neuropathy, baroreflex failure

This is your diagnostic compass.

Section 7—Putting It All Together: Cross-Test Integration

The framework becomes exponentially more powerful when combined with individual test domains.

Integrated Interpretation Table

Baseline	Deep Breathing	Valsalva	Standing	Pattern	Meaning
Normal	Normal	Normal	Normal	NN	Intact ANS
Normal	Abnormal	Abnormal	Normal	NA	Early mixed dysfunction
Normal	Normal	Abnormal	Abnormal	NA	Baroreflex-dominant issue
Abnormal	Abnormal	Abnormal	Abnormal	AA	Advanced autonomic failure
Normal	Abnormal	Normal	Abnormal	NA	Parasympathetic injury
Abnormal	Normal	Abnormal	Abnormal	AA	Sympathetic + baroreflex failure

Integrating tests creates mechanistic clarity—instantly.

Science Snapshot—The Three-Pattern Autonomic Model

Three Categories → Three Diagnoses → Three Pathways

Category	Mechanism	Typical Diagnosis
Normal-Normal	Intact	Normal ARS*, episodic OI
Normal-Abnormal	Impaired reserve	POTS, early neuropathy
Abnormal-Abnormal	Global failure	OH, baroreflex failure

*ARS = Autonomic Reflex Screen.

This pattern model is the shortest route to interpretation accuracy.

Translation—What This Chapter Really Means

- "Normal" doesn't mean symptoms aren't real—it means the test didn't push the right buttons.
- Many dysautonomia patients fall into the "Normal-Abnormal" category, where resting function looks fine but compensation collapses.
- "Abnormal-Abnormal" signals advanced or widespread dysfunction.
- This framework eliminates confusion and clarifies the mechanism instantly.
- If a clinician doesn't use this system, they are reading tests in low resolution.

Tilt Tip—How to Use This Framework to Advocate for Yourself

- If your tests are Normal-Abnormal, emphasize that your symptoms appear under physiologic stress—not at rest.

- If all results are normal yet symptoms persist, push for heat, exercise, or prolonged upright testing.
- If your results are Abnormal-Abnormal, advocate for targeted autonomic therapy, not generic reassurance.
- Use the pattern classification to steer clinicians toward mechanism-based treatment—not guesswork.

CHAPTER 11
MEDICATION INTERFERENCE: WHEN PHARMACOLOGY WEARS A FAKE MUSTACHE

Because nothing ruins autonomic testing faster than a medication pretending to be your nervous system.

Introduction—The Pharmacologic Identity Theft Problem

Autonomic testing is supposed to reveal the truth about your nervous system.

Medications, however, often show up wearing disguises: beta-blockers pretending to be your vagus nerve, stimulants impersonating sympathetic tone, antidepressants sabotaging baroreflexes, diuretics manufacturing hypotension, and benzodiazepines cosplaying as parasympathetic excellence.

This chapter exposes how medications distort test results, confuse interpretation, and create false patterns that mimic or mask dysautonomia.

When pharmacology steps in, physiology steps aside—and you're left reading a story written by the medication, not the nervous system.

Welcome to autonomic testing in the era of fake mustaches.

Section 1—Why Medication Interference Matters More Than Clinicians Admit

Autonomic testing relies on reflex timing, amplitude, compensation, reserve, and physiologic adaptability.

Medications modify neurotransmitters, receptor sensitivity, heart rate, vascular tone, adrenal output, sweating patterns, and baroreflex behavior.

In other words, drugs alter every mechanism the tests rely on. If you don't know what the medication is doing, you don't know what the test is saying.

Section 2—The Medication Interference Map

Every drug that touches the autonomic system falls into one of five categories:

1. Maskers—make dysfunction look normal.
2. Exaggerators—make normal physiology look abnormal.
3. Hijackers—replace neural responses entirely.
4. Saboteurs—create abnormal patterns from scratch.
5. Shape-shifters—produce inconsistent, unpredictable changes.

These categories form the backbone of interpretation.

Section 3—The Maskers (Drugs That Make Abnormal Look Normal)

Some medications create artificially "good-looking" results. These drugs hide dysfunction by propping up tone that your nervous system cannot generate.

Category: Maskers

Medication	Effect	Misleading Appearance
Beta-blockers	Reduce HR	Normal HR response despite POTS
Benzodiazepines	Enhance vagal tone	Strong RSA despite parasympathetic impairment
SSRIs/SNRIs	Stabilize HR variability	Artificially smoothed curves
Fludrocortisone	Increases blood volume	Improved orthostatic BP falsely normalized
Midodrine	Enhances vasoconstriction	Masks sympathetic underactivation

Maskers are particularly mischievous because they hide mechanisms.

A patient on midodrine, beta-blockers, and fludrocortisone—looks entirely different off medication.

Maskers create false reassurance.

Pyridostigmine (Mestinon): The Parasympathetic Polisher

Pyridostigmine deserves a quiet spotlight here. By inhibiting acetylcholinesterase, it boosts parasympathetic tone just enough to make weak RSA, shallow Valsalva braking, or borderline orthostatic compensation look more respectable than they actually are. It doesn't fix the mechanism—it just props it up.

Testing on Mestinon often underestimates the true severity of vagal and baroreflex impairment, because what you're seeing is a medicated performance, not the patient's unassisted physiology.

Section 4—The Exaggerators (Drugs That Make Normal Look Pathologic)

Exaggerators amplify physiologic responses, mimicking hyperadrenergic states or severe dysautonomia.

Category: Exaggerators

Medication	Effect	False Pattern
Stimulants (amphetamine, methylphenidate)	Increase NE	Hyperadrenergic POTS patterns
Decongestants (pseudoephedrine)	Vasoconstriction + HR↑	Exaggerated Phase II responses
Levothyroxine (over-replaced)	Increased metabolic drive	Unstable HR curves
Caffeine	Acute SNS boost	Tremulous HR, spiked curves
SNRIs	Elevated NE	Sympathetic contamination

These drugs produce jittery HR morphology, adrenaline-like curves, and unstable BP traces. That is not dysautonomia. That is pharmacology shouting through a megaphone.

Section 5—The Hijackers (Drugs That Replace Normal Physiology)

These medications don't just modify responses—they take over the entire pathway and pretend to be your nervous system.

Category: Hijackers

Medication	Effect	Impact on Testing
Anticholinergics	Block vagal pathways	No deep breathing RSA
TCAs	Suppress parasympathetic activity	False parasympathetic failure
Alpha-2 agonists (clonidine, guanfacine)	Blunt SNS tone	Weak Phase II Valsalva
Nitrates	Forced vasodilation	Collapse of BP responses
SGLT2 inhibitors + diuretics	Reduce volume	Orthostatic hypotension mimic

These drugs override the signals that autonomic tests attempt to measure. The result: you end up testing the medication—not the nervous system.

Section 6—The Saboteurs (Drugs That Produce Abnormality Out of Thin Air)

These medications create dysfunction even when the autonomic system is totally healthy.

Category: Saboteurs

Medication	Effect	False Signal
Diuretics	Hypovolemia	Orthostatic hypotension
GLP-1 agonists	Slowed gastric emptying	GI-related vagal noise
Antihypertensives	Vascular relaxation	Blunted BP responses
Alcohol (acute)	Vagal suppression	Low HRV
Cannabinoids	Parasympathetic distortion	Unpredictable curves

Saboteurs generate abnormalities that resemble neuropathy, baroreflex issues, poor compensation, or vagal collapse. But the culprit is pharmacologic—not physiologic.

Section 7—The Shape-Shifters (The Most Unpredictable Class)

Shape-shifters produce erratic, inconsistent autonomic responses that vary test-by-test, making interpretation nearly impossible.

Category: Shape-Shifters

Medication	Why They Are Unpredictable
Antipsychotics	Mixed receptor activity (alpha, muscarinic, histamine)
Tricyclics	Anticholinergic + NE reuptake blockade
Anticonvulsants	Central sedation + peripheral tone changes
Opioids	Central vagal enhancement + peripheral vasodilation
Hormone therapy	Alters vascular reactivity

Shape-shifters produce inconsistencies that mimic patchy neuropathy, episodic hyperadrenergic states, presyncope, and baroreflex chaos. Their effects cannot be interpreted cleanly without knowing the drug list.

Section 8—Test-by-Test: How Medications Alter Each Autonomic Domain

Resting State Distortion

Drug Type	Effect
Stimulants	Elevated HR + poor HRV
Beta-blockers	Artificially low HR
Sedatives	Smoothed HR patterns
Vasodilators	Low BP

Baseline distortions are often the most misleading.

Deep Breathing Distortion

Drug Effect	Result
Anticholinergic	No RSA (false parasympathetic failure)
Benzodiazepine	Exaggerated RSA (fake parasympathetic strength)
Beta-blockers	Reduced peak HR (confused interpretation)

Deep breathing is the most medication-sensitive test.

Valsalva Distortion

Drug Type	Impact on Phase II	Impact on Phase IV
Alpha agonists	Overshoot suppression	Weak brake
Stimulants	Exaggerated responses	Chaotic recovery
Clonidine	Blunted compensation	Flat overshoot
Nitrates	Collapse of Phase II	No Phase IV

Valsalva is the most distorted by vasodilators.

Standing/Tilt Distortion

Drug Type	Orthostatic Impact
Diuretics	OH mimic
Stimulants	POTS-like tachycardia
Beta-blockers	Hidden tachycardia mechanisms
Midodrine	Masked sympathetic dysfunction

Upright patterns become unreadable without drug context.

Section 9—How to Interpret Tests Accurately When Medications Are Involved

Principle 1—Assume Nothing
If the patient is on ANY medication that affects tone, every abnormality must be cross-checked against the drug list.

Principle 2—Identify the Category (Masker, Exaggerator, Hijacker, Saboteur, Shape-Shifter)
Determine which distortion is likely.

Principle 3—Look for Mechanisms That Don't Fit
If the pattern is too clean—or too chaotic—consider medication effects.

Principle 4—Integrate Across Domains
If only one test is abnormal, suspect medications. If multiple domains are abnormal in a consistent pattern, suspect physiology.

Principle 5—Look at Morphology Over Magnitude
Drugs often modify amplitude but rarely reproduce precise timing signatures of true dysfunction.

Science Snapshot—Medication Interference Summary Table

Interference Type	Mechanism	What It Fools You Into Thinking
Masker	Compensates for failure	Everything looks normal
Exaggerator	Amplifies SNS tone	Hyperadrenergic POTS
Hijacker	Blocks reflex pathways	Severe neuropathy
Saboteur	Creates dysfunction artificially	Orthostatic hypotension
Shape-Shifter	Inconsistent effects	Baroreflex chaos

This table alone prevents 50% of misinterpretations.

Translation—What This Chapter Really Means

- Medications distort autonomic testing more than any other factor.
- Some hide dysfunction; others create it.
- A normal test on medication does *not* mean the autonomic system is normal.
- An abnormal test on medication does *not* guarantee pathology.
- Mechanism must be interpreted through the lens of pharmacologic effects.
- If you ignore medications, you are diagnosing shadows.

Tilt Tip—How to Apply This in Real Life

- Bring your full medication list when undergoing testing—including supplements.
- If possible, testing should be done before starting autonomic-active medications.

- If testing must be done on medication, interpretation should focus on patterns, not magnitudes.
- Use the category system (Masker, Exaggerator, Hijacker, Saboteur, Shape-Shifter) to understand how each drug influences results.
- When your clinician dismisses abnormalities, remind them, "The medication may be interfering with the reflex—not the symptom."

CHAPTER 12
HEAT, FOOD, STRESS, AND STANDING IN THE REAL WORLD: FUNCTIONAL TRIGGERS

Because dysautonomia doesn't happen in a vacuum—it happens in public, usually at the worst possible time.

Introduction—When Real Life Becomes a Physiologic Obstacle Course

Autonomic testing is performed in a peaceful, climate-controlled room where nothing bad ever happens.

Real life, on the other hand, involves sunlight, gravity, meals, unpredictable emotions, surprise deadlines, humidity, crowded grocery stores, and absolutely no respect for your autonomic reserve.

This chapter covers the four real-world trigger categories that reliably destabilize dysautonomia physiology: heat, food, stress, and standing.

Each trigger interacts with autonomic pathways in specific, predictable, mechanistic ways—none of which are psychological, dramatic, or optional.

These are physiologic landmines. You step on one → your ANS detonates.

Section 1—Heat: The Most Ruthless Trigger of Them All

Heat is the autonomic system's worst enemy, primarily because it expands blood vessels, amplifies pooling, suppresses sympathetic effectiveness, and demands increased cardiac output.

Heat exposure forces the ANS to dilate skin vessels, move blood toward the surface, increase sweating, increase heart rate, and maintain organ perfusion.

A healthy nervous system handles this with grace. A dysautonomic nervous system handles this with collapse.

Why Heat Is Devastating in Dysautonomia

Heat Effect	Mechanism	Consequence
Vasodilation	Pooling increases	Low preload
Sweating demand	Sudomotor burden	Autonomic fatigue
Increased HR	Compensatory	Tachycardia
Increased metabolic rate	Higher O_2 demand	Immediate fatigue
Blood redistribution	Less to brain	Dizziness, fog

Heat intolerance is not "being sensitive." It is physics + impaired physiology.

Heat-Induced Symptom Patterns: racing heart, flushing, sweating (or impaired sweating), nausea, disorientation, cognitive shutdown, and a sense of impending doom (physiologic, not psychological).

Heat Failure Curve: Predictable Timing

Exposure Duration	Physiologic Impact
1–3 minutes	HR increases
3–7 minutes	BP destabilizes
7–15 minutes	Stroke volume drops
>15 minutes	Cognitive fog, fatigue, presyncope

This is predictable because the mechanisms are predictable.

Section 2—Food: The Splanchnic Blood Donation You Didn't Volunteer For

Food does more than provide calories. It triggers one of the largest vascular redistribution events outside of exercise.

After you eat, blood shifts to the gut (postprandial pooling), vascular tone adjusts, vagal pathways activate, sympathetic regulation quiets, glucose levels peak, insulin surges, and gastrointestinal motility triggers sensory pathways.

If the autonomic system cannot compensate, everything spirals.

Why Meals Trigger Symptoms

Food Effect	Mechanism	Outcome
Blood shifts to the gut	Splanchnic dilation	Low preload
Vagal activation	HR↓ in some, HR chaos in others	Instability
Rapid carb absorption	Glucose spike	Fatigue + catecholamine turbulence

Food Effect	Mechanism	Outcome
Thermal effect of food	Increased metabolic demand	Tachycardia
GI motility	Vagal reflex	Nausea, dizziness

Meals are mini stress tests—three times a day, every day.

Postprandial Symptom Pattern: worsening POTS symptoms, increased lightheadedness, elevated HR, brain fog, heavy limbs, fatigue, and GI distress.

Carbohydrates amplify the effect more than protein or fat. Large meals magnify the dysfunction. Warm meals are even worse (hello, soup-triggered dizziness).

High-Risk Meal Features

Meal Type	Physiologic Burden
Large meals	Maximal pooling
High-carb	Glucose + insulin surge
Warm meals	Thermogenic load
Rapid eating	Excessive vagal activation
Greasy meals	Delayed gastric emptying

The autonomic system must compensate for all of this—often unsuccessfully.

Section 3—Stress: The Biological Drama You Cannot Avoid

Stress is not emotional fragility. Stress is a physiologic event involving adrenaline, cortisol, glucose mobilization, vasoconstriction, HR elevation, and a shift in autonomic balance. The ANS interprets all stress—emotional, cognitive, or physical—as threat physiology.

Stress Mechanisms That Wreck Dysautonomia

Stress Response	Mechanism	Impact
NE release	Sympathetic surge	HR↑, tremor
Cortisol	Shifts glucose metabolism	Fatigue
Vasoconstriction	Maldistributed tone	BP inconsistency
Baroreflex disruption	Rhythmic instability	Dizziness
Hyperventilation tendency	CO_2 drop	Cognitive fog

Dysautonomia is not incompatible with stress—it's demolished by it.

Stress-Induced Symptom Patterns: trembling, tachycardia, BP swings, chest tightness, cognitive shutdown, sensory overload, and fatigue afterward ("post-adrenal crash").

This is not anxiety disorder. This is inadequate autonomic buffering.

Stress Load Timeline

Timing	Physiologic Stage
0–1 minute	Adrenaline spike
1–5 minutes	HR + BP fluctuations

Timing	Physiologic Stage
5–20 minutes	Cortisol wave
20–60 minutes	Post-adrenal fatigue
Hours later	Autonomic exhaustion

Stress leaves a metabolic crater behind.

Section 4—Standing: The Trigger Guaranteed to Find You Daily

Standing is the most predictable trigger because gravity is consistent and merciless.

Standing causes blood pooling, reduced venous return, increased HR, and baroreflex engagement. If the system fails to compensate, symptoms bloom within seconds.

Why Standing Affects Autonomic Reserve

Upright Effect	Mechanism	Impact
500–1000 mL blood shifts downward	Venous pooling	Low cardiac output
Baroreflex activation	Tries to stabilize BP	May fail
HR increase	Compensatory	May overshoot
Cerebral perfusion ↓	Oxygenation drop	Dizziness, fog

Standing is a physiologic interrogation. The ANS either answers correctly —or collapses.

Standing Symptom Patterns: rapid HR increase, dizziness, heat sensation, blurred vision, tremor, weakness, nausea, and chest tightness.

Symptoms vary by mechanism:
- Neuropathic → weak vasoconstriction
- Hyperadrenergic → excessive HR, flushing
- Hypovolemic → narrow pulse pressure

Standing is not optional. Symptoms therefore rarely retreat.

Section 5—Trigger Stacking: The Real-World Disaster Scenario

The biggest issue is not each trigger individually. It is when they combine.

Common Stacking Scenarios:

1. Heat + Standing
   - The classic outdoor collapse

2. Food + Standing
   - Grocery store checkout line meltdown

3. Stress + Heat
   - Workplace meeting in a warm room

4. Food + Heat
   - Restaurants with warm ambiance and warm meals = physiologic horror

5. Standing + Stress
   - Public speaking
   - Waiting in line
   - Social events

6. Food + Stress + Standing
- Weddings
- Conferences
- Family gatherings
- Airports (the Olympics of trigger stacking)

The autonomic system has limits. Stacking pushes it past them quickly.

Section 6—Mechanism-Based Trigger Interpretation

Trigger → Physiologic Failure → Clinical Interpretation

Trigger	Mechanism	Failure Mode
Heat	Vasodilation	Preload drops
Food	Splanchnic pooling	Baroreflex + vagal conflict
Stress	Sympathetic surge	Rhythm instability
Standing	Gravity + pooling	Orthostatic intolerance

Triggers are diagnostic tools in disguise.

Section 7—How to Identify Which Trigger Is Dominant

Symptoms vary by trigger dominance:

1. Heat-Dominant Dysautonomia
   - Flushing
   - Palpitations
   - Fatigue
   - Collapse in warm rooms

Mechanism: vasodilation + sudomotor failure.

2. Food-Dominant Dysautonomia
- Postprandial dizziness
- Fog
- Tachycardia after meals

Mechanism: splanchnic pooling + vagal conflict.

3. Stress-Dominant Dysautonomia
- Tremor
- Racing heart
- Sensory overload

Mechanism: sympathetic noise + cortisol.

4. Standing-Dominant Dysautonomia
- Instant dizziness
- Tachycardia
- Narrow pulse pressure

Mechanism: preload failure under gravity.

Patterns do not lie.

Science Snapshot—Triggers and Autonomic Pathways

Trigger	Branch Involved	Mechanism	Outcome
Heat	Sympathetic cholinergic	Vasodilation	Pooling
Food	Vagal + sympathetic	Splanchnic redistribution	Hypotension
Stress	Sympathetic	NE + cortisol	Instability

Trigger	Branch Involved	Mechanism	Outcome
Standing	Sympathetic	Vascular compensation	OI or POTS

Understanding triggers clarifies mechanisms better than most tests.

Translation—What This Chapter Really Means

- Heat is a physiologic wrecking ball.
- Food steals blood from your heart and gives it to your intestines.
- Stress hijacks the autonomic system in seconds.
- Standing is the ultimate daily enemy.
- Symptoms are predictable, mechanistic, and measurable—not psychological.
- Trigger stacking creates the worst episodes.
- Recognizing your dominant trigger is the key to managing real-world crashes.

Tilt Tip—How to Use Trigger Data in Daily Life

- Avoid heat—your autonomic system doesn't negotiate with it.
- Eat smaller, cooler, lower-carb meals to minimize postprandial collapse.
- Build stress-buffering strategies before—not after—your system breaks.
- Use compression garments and counter-maneuvers for standing triggers.
- Never stack triggers intentionally—your ANS is already outnumbered.
- Track which triggers affect you the most—this determines your real-world strategy.

PART IV
MECHANISMS EXPOSED: IDENTIFY YOUR ENEMY

Where physiology finally confesses.

CHAPTER 13
TEST INTEGRATION: IDENTIFYING THE DOMINANT FAILURE MODE

Because knowing each test result is cute, but knowing how they work together is how you actually win the war.

Introduction—Single Tests Are Clues; Integration Is the Confession

Every autonomic test—baseline, deep breathing, Valsalva, standing, sudomotor—gives you a piece of the puzzle. Individually, they whisper. Together, they shout.

Patients don't present with isolated mechanisms. They present with combinations—venous pooling plus baroreflex delays, vagal impairment plus sympathetic noise, heat intolerance plus postprandial chaos, and neuropathy plus orthostatic collapse.

If you want to understand the actual physiologic failure—not the symptom list—you must integrate every test result into a single mechanistic narrative.

This chapter shows you how to identify the dominant failure mode: the one mechanism that explains the entire pattern. Once you identify the dominant failure mode, the rest of the diagnosis becomes inevitable.

Section 1—What "Dominant Failure Mode" Means

A failure mode is the primary physiologic mechanism responsible for the clinical pattern.

In dysautonomia, this is usually one of the following:
1. Preload failure (pooler physiology)
2. Sympathetic underactivation
3. Sympathetic overactivation (compensatory or primary)
4. Parasympathetic impairment
5. Baroreflex timing dysfunction
6. Small Fiber Neuropathy (sudomotor-driven)
7. Combined failure (two or more systems refusing to cooperate)

Every patient has one dominant mechanism—even if several are involved. Identifying it is the difference between treating symptoms and actually treating physiology.

Section 2—Step 1: Start With the Resting State (The Baseline Truth Serum)

Resting physiology tells you where the system starts before being stressed.

Baseline Interpretation Matrix

Finding	Meaning	Dominant Failure Mode Suggestion
High HR	Preload failure	Pooler physiology
Low HRV	Parasympathetic impairment	Vagal weakness
Labile BP	Baroreflex timing issue	Baroreflex dysfunction
Wide PP	Vasodilation	Heat sensitivity
Narrow PP	Low stroke volume	Preload failure
Tremulous HR	Sympathetic noise	Hyperadrenergic state

Baseline abnormalities already hint at the mechanism, but they are not definitive. You need to stress the system to reveal the truth.

Section 3—Step 2: Deep Breathing (The Parasympathetic Polygraph)

Deep breathing is the first targeted interrogation.

Deep Breathing Interpretation Table

Finding	Meaning	Dominant Failure Mode
Low amplitude	Vagal impairment	Parasympathetic failure
Delayed trough	Baroreflex timing issue	Central dysfunction
Irregular oscillations	Sympathetic contamination	Hyperadrenergic state
Flat line	Severe vagal failure	Parasympathetic collapse

Deep breathing identifies parasympathetic reserve—one of the earliest autonomic failure signals.

Section 4—Step 3: Valsalva (The Full-System Stress Scenario)

Valsalva reveals sympathetic power, baroreflex timing, and vagal braking in one test.

Phase Interpretation

Abnormal Phase	Mechanism Failure	Dominant Mode
Phase II weak	Sympathetic underactivation	Neuropathic/Ganglionic
Phase II spike	Sympathetic excess	Hyperadrenergic

Abnormal Phase	Mechanism Failure	Dominant Mode
Phase IV absent	Baroreflex failure	Central or advanced neuropathy
Phase IV delayed	Timing issue	Baroreflex dysfunction
HR brake poor	Parasympathetic impairment	Vagal failure

Valsalva is the most potent integrator because it divides dysfunction into precise moments.

Section 5—Step 4: Standing/Tilt (The Real-World Battlefield)

Nothing reveals the dominant failure mode faster than upright physiology.

Standing Interpretation Table

Pattern	Mechanism	Dominant Failure Mode
HR↑ ≥30 with BP stable	Preload failure	POTS phenotype
HR↑ + PP↓	Low stroke volume	Preload failure
BP↓ early	Sympathetic underactivation	OH
BP↓ late	Baroreflex fatigue	Delayed OH
HR↑ + tremor + heat	Sympathetic excess	Hyperadrenergic
HR↓ + BP↓	Vagal surge	Vasovagal
BP oscillation	Baroreflex instability	Central dysfunction

Standing physiology reveals who cannot maintain volume, who cannot generate tone, who overcompensates, and who misfires altogether.

Section 6—Step 5: Sudomotor (The Early Warning System)

Sudomotor abnormalities clarify where the dysfunction originates.

Sudomotor Interpretation

Pattern	Meaning	Dominant Failure Mode
Distal loss	Length-dependent neuropathy	Neuropathic dysautonomia
Patchy loss	Immune-mediated injury	Autoimmune
Global low output	Severe sympathetic failure	Autonomic neuropathy
Hyper-conductance	Sympathetic excess	Hyperadrenergic
Asymmetry	Focal nerve dysfunction	Segmental neuropathy

Sudomotor is often the first test to show neuropathic patterns—even before cardiac reflexes.

Section 7—Timing and Coordination Failures: When Reflexes Don't Agree on the Script

Not all autonomic dysfunction is about strength or weakness. Some of the worst physiology comes from timing errors, where components of the reflex arc are technically intact but fire out of sequence. These patterns often masquerade as normal until the moment the system falls apart.

A timing failure means the autonomic system recognizes the problem too late, responds too aggressively, or corrects in a way that destabilizes the very compensation it was supposed to support. The autonomic branches are not weak—they are disorganized. This is where the most dramatic failure modes live: physiology that behaves consistently until it suddenly doesn't.

Paradoxic Parasympathetic Syndrome (PPS): When the Brake Fires at the Worst Possible Time

PPS is one of the clearest timing failures in autonomic medicine. It is not a sign of high vagal tone or emotional reactivity. PPS is a delayed, exaggerated parasympathetic correction triggered only after the sympathetic system has already strained to maintain perfusion. The system does not lose the ability to respond—it loses the ability to respond on time.

Mechanism Breakdown:
1. Stroke volume fails.
2. Sympathetic activation rises appropriately.
3. The baroreflex detects the crisis late.
4. The parasympathetic system overcorrects.
5. Heart rate and blood pressure collapse abruptly.

PPS Across Reflex Testing

Reflex Test	PPS Pattern	Interpretation/Meaning
Deep Breathing	Normal pattern at rest	PPS is not baseline vagal weakness; dysfunction emerges only under baroreflex stress
Valsalva	Weak/Delayed Phase II; exaggerated Phase IV	Delayed baroreflex recognition with oversized parasympathetic correction
Standing/Tilt	Initially stable, then abrupt HR/BP drop	Classic PPS: prolonged sympathetic compensation followed by delayed vagal surge

Why PPS matters: PPS is a timing disorder, not a tone disorder. It predicts collapses that appear suddenly, unprovoked, and inconsistent with resting physiology.

PPS Diagnostic Pattern Table

Component	PPS Pattern	Interpretation
Phase II (Valsalva)	Weak, delayed, inconsistent	Sympathetic struggle to stabilize pressure
Phase IV (Valsalva)	Oversized overshoot	Baroreflex overcorrection
Standing HR	Rises, then abruptly falls	Vagal overshoot triggered by delayed reflex
Standing BP	Drops suddenly after compensation	Loss of vascular tone + bradycardia
HRV	Often normal	PPS is timing-based, not baseline-based
Symptoms	Sudden collapse	Timing failure

Delayed baroreflex recognition → exaggerated parasympathetic overcorrection → collapse.

Your vagus nerve shows up late, panics, and hits the brake too hard. Interpret PPS by looking at timing, not magnitude. PPS is identified by *when* the system fails, not just *how*.

Section 8—Integrating All Tests: The Mechanistic Algorithms

Algorithm 1—Preload Failure (Pooling Physiology)
Clues From Each Test

Test	Finding
Baseline	High HR, narrow PP
Deep Breathing	Normal or mildly low
Valsalva	Strong HR rise, weak BP stabilization
Standing	HR↑ ≥ 30–40 bpm, PP↓, BP stable or variable
Sudomotor	Often normal or mildly abnormal

Dominant Failure Mode: Preload Failure → POTS Phenotype → Reduced Stroke Volume

Mechanism: Not enough blood returning to the heart.

Algorithm 2—Sympathetic Underactivation (Neuropathic Pattern)
Clues From Each Test

Test	Finding
Baseline	Normal HR or mild tachycardia
Deep Breathing	Normal or weak
Valsalva	Weak Phase II, no stabilization
Standing	Early BP drop
Sudomotor	Distal or global loss

Dominant Failure Mode: Sympathetic Underactivation → Orthostatic Hypotension

Mechanism: Nerves cannot constrict vessels.

Algorithm 3—Sympathetic Excess (Hyperadrenergic Pattern)
Clues From Each Test

Test	Finding
Baseline	Tremor, high HR, high PP, sympathetic noise
Deep Breathing	Low amplitude but noisy
Valsalva	Exaggerated HR responses
Standing	HR↑ + BP↑ or stable but chaotic
Sudomotor	Hyper-conductance

Dominant Failure Mode: Sympathetic Overactivation → Hyperadrenergic POTS

Mechanism: Overactive NE release → tone imbalance.

Algorithm 4—Parasympathetic Impairment
Clues From Each Test

Test	Finding
Baseline	Low HRV
Deep Breathing	Low or absent RSA
Valsalva	Poor HR braking
Standing	High HR variability + instability
Sudomotor	May be normal or patchy

Dominant Failure Mode: Vagal Failure → Reduced Modulation

Mechanism: The braking system doesn't brake.

Algorithm 5—Baroreflex Timing Dysfunction
Clues From Each Test

Test	Finding
Baseline	Labile BP
Deep Breathing	Delayed trough
Valsalva	Phase transitions mistimed
Standing	BP oscillation
Sudomotor	Varied

Dominant Failure Mode: Baroreflex Timing Error

Mechanism: The reflex arc responds, but late or inconsistently.

Algorithm 6—Small Fiber Neuropathy (Postganglionic)
Clues From Each Test

Test	Finding
Baseline	Variable
Deep Breathing	Inconsistent
Valsalva	Often mixed weak responses
Standing	Mild–moderate OI
Sudomotor	Patchy or distal low output

Dominant Failure Mode: Small Fiber Neuropathy → Patchy Autonomic Failure

Mechanism: Nerve endings do not relay signals.

Algorithm 7—Combined Failure
When two or more systems collapse simultaneously, patterns look chaotic—but integration clarifies them.

Typical combined patterns include:
- Vagal weakness + preload failure
- Neuropathy + baroreflex delay
- Hyperadrenergic compensation + low stroke volume

Combined failure modes are still analyzable as long as you identify the dominant initiating mechanism.

Section 9—The Integration Table (The Autonomic Master Map)
Integration Matrix

Baseline	Deep Breathing	Valsalva	Standing	Sudomotor	Dominant Failure Mode
HR↑, PP↓	Normal	Strong spike	HR↑ ≥ 30, PP↓	Normal	Preload failure
Normal	Low amplitude	Poor brake	Unstable HR	Normal	Vagal impairment
Normal	Normal	Weak Phase II	BP↓	Distal loss	Sympathetic under-activation
Tremor	Noisy	Exaggerated	HR↑ + BP↑	Hyper	Hyper-adrenergic

Baseline	Deep Breathing	Valsalva	Standing	Sudomotor	Dominant Failure Mode
Labile BP	Delayed	Delayed	Oscillation	Variable	Baroreflex timing
Normal	Mild low	Mixed	Mild OI	Patchy	Small fiber neuropathy
Abnormal	Abnormal	Abnormal	Abnormal	Abnormal	Global autonomic failure

Interpretation becomes shockingly simple.

Science Snapshot—The Dominant Failure Mode Quick Guide

Failure Mode	Core Problem	Classic Pattern
Preload failure	Low stroke volume	POTS
Sympathetic underactivation	Weak vasoconstriction	OH
Sympathetic excess	NE overdrive	Hyper-POTS
Vagal impairment	Poor HR modulation	Low RSA + poor braking
Baroreflex dysfunction	Timing errors	BP oscillation
Small fiber neuropathy	Nerve loss	Distal sweat loss
Global failure	Multisystem abnormal-abnormal	Varies
Paradoxic Parasympathetic Syndrome (PPS)	Delayed baroreflex detection → exaggerated vagal overshoot	Weak Phase II, oversized Phase IV, sudden HR/BP collapse on tilt

Translation—What This Chapter Really Means

- Autonomic testing is only powerful when the results are integrated, not isolated.
- Every patient has a dominant failure mode driving symptoms.
- Recognizing the primary mechanism transforms treatment from generic to targeted.
- The integration framework reveals whether dysfunction is preload-driven, neuropathic, hyperadrenergic, vagal, baroreflex, or mixed.
- Mechanism, not symptoms, should guide management.

Tilt Tip—How to Use This in Real Life

- Ask your clinician which dominant failure mode your results point to.
- If they can't answer, request integrated interpretation—one domain is not enough.
- Track patterns: heat, meals, stress, and standing all correlate with specific mechanisms.
- Use your failure mode to select targeted treatments (covered in later chapters).
- Base your management plan on physiology, not guesswork.

CHAPTER 14
PROGNOSTIC INDICATORS: WHAT PREDICTS PROGRESSION VS. RECOVERY

Because everyone wants to know the future—even when their autonomic system is still trying to survive the present.

Introduction—Prognosis Is Not a Guess; It's a Pattern

People ask, "Is my dysautonomia going to get better, worse, or stay the same?" Clinicians often respond with vague optimism, unjustified pessimism, or the ever-reliable "it depends." But here's the truth—the ANS leaves clues. Prognosis is predictable when you know what to look for.

Your autonomic tests, symptom triggers, recovery patterns, stress tolerance, heat response, and sudomotor findings create a recognizable trajectory—one that points toward recovery, chronic stability, or progression.

This chapter teaches you how to read those trajectories like a seasoned autonomic specialist—without sugarcoating, catastrophizing, or relying on hopeful guesses.

Section 1—Prognosis Begins With Mechanism, Not Symptoms

Symptoms are noisy. Mechanisms are reliable.

The Mechanism–Prognosis Rule

Mechanism	General Prognosis
Preload failure	Highly treatable
Hyperadrenergic	Variable but often improvable
Vagal impairment	Slow recovery, but possible
Baroreflex dysfunction	Variable; depends on timing accuracy
Small fiber neuropathy	Depends on etiology
Global autonomic failure	Least reversible but improvable in function

Symptoms do not dictate outcome—mechanism does.

Section 2—The Three Prognostic Classes

Autonomic disorders follow three predictable trajectories:

1. Improve-ready physiology (the recoverable group)
These individuals have mechanisms that respond well to targeted therapy. Typical features include preload failure, hyperadrenergic compensation, mild vagal impairment, and early small fiber involvement.

2. Stable-but-struggling physiology (the chronic management group)
These individuals may not improve dramatically, but they often achieve long-term stability with the right strategies. Typical features include moderate neuropathy, baroreflex timing delays, and combined vagal and sympathetic involvement.

3. Progressive physiologic injury (the high-monitoring group)
This group shows multi-domain dysfunction that worsens without targeted intervention. Typical features include severe autonomic neuropathy, autoimmune ganglionic injury, and degenerative processes.

Prognosis is not one-size-fits-all—it's mechanistic.

Section 3—The 10 Most Powerful Prognostic Indicators

These markers predict recovery potential far more accurately than symptom descriptions.

Indicator 1—Baseline Stability

A stable resting state (HR, BP, HRV) indicates the nervous system retains enough reserve to adapt and recover.

Prognostic Interpretation

Baseline Pattern	Prognosis
Stable HR/BP	Strong recovery potential
Mild instability	Responsive to treatment
High HR + low HRV	Slow improvement
BP lability	Baroreflex-driven chronicity
Chaotic baseline	Multi-domain involvement

Indicator 2—Deep Breathing Amplitude

This measures parasympathetic reserve.

Interpretation

RSA Amplitude	Meaning	Prognosis
Strong	Robust vagal tone	Excellent
Moderate	Mild impairment	Good
Low	Vagal weakness	Slower progress
Absent	Severe impairment	Guarded

Recovery depends on how much braking power the vagus nerve can generate.

Indicator 3—Valsalva Phase II Strength

Sympathetic activation ability is a powerful predictor of upright success.

Phase II Response	Mechanism	Prognosis
Strong stabilization	Good vasoconstriction	Excellent
Moderate	Partial ability	Good
Weak	Sympathetic failure	Depends on etiology
Absent	Advanced dysfunction	Guarded

Valsalva reveals whether the sympathetic system can fight for you.

Indicator 4—Valsalva Phase IV Overshoot

This is baroreflex magic. If present, your reflex arc works. If absent, the reflex arc is malfunctioning.

Prognosis by Overshoot Pattern

Overshoot	Meaning	Prognosis
Normal	Intact baroreflex	Excellent
Weak	Mild timing issue	Good

Prognostic Indicators

Overshoot	Meaning	Prognosis
Delayed	Central timing error	Variable
Absent	Baroreflex failure	Guarded

Indicator 5—Standing Pulse Pressure

Pulse pressure (PP) predicts stroke volume and preload.

PP Pattern	Interpretation	Prognosis
Stable	Adequate stroke volume	Excellent
Mild narrowing	Treatable preload issue	Good
Severe narrowing	Low SV	Depends on response to therapy
Collapse	Severe orthostatic intolerance	Guarded

Improving PP is one of the strongest predictors of recovery.

Indicator 6—Sudomotor Integrity

Sweat gland patterns reveal small fiber health.

Sudomotor Pattern	Prognosis
Normal	Excellent recovery potential
Mild distal loss	Good with treatment
Patchy loss	Variable; depends on cause
Global failure	Guarded unless reversible cause

Sudomotor results track progression better than HR or BP.

Indicator 7—Heat Tolerance

Heat intolerance correlates with autonomic reserve.

Heat Response	Prognosis
Tolerable	Strong reserve
Moderate intolerance	Improvable
Severe intolerance	Low vascular reserve
Heat collapse	Significant impairment

Better heat tolerance → better prognosis.

Indicator 8—Postprandial Stability

If meals overwhelm you, your autonomic system is fighting splanchnic pooling poorly.

Response to Meals	Mechanism	Prognosis
Stable	Good redistribution	Excellent
Mild dizziness	Treatable	Good
Significant OI	Poor splanchnic compensation	Variable
Collapse	Severe autonomic stress	Guarded

Postprandial physiology predicts daily functionality.

Indicator 9—Stress Sensitivity

If stress dramatically destabilizes physiology, prognosis depends on sympathetic modulation capacity.

Stress Response	Meaning	Prognosis
Manageable	Good buffering	Excellent
Moderate	Hyperadrenergic	Good/Variable

Stress Response	Meaning	Prognosis
Severe	Baroreflex disruption	Guarded
Shutdown	Multi-domain failure	Guarded

Indicator 10—Response to Interventions

This is the strongest predictor. If the patient responds to hydration, compression, salt loading, beta-blockers, midodrine, fludrocortisone, or HCT*/POTS-specific exercise, the prognosis is significantly better.

HCT = Horizontal Conditioning Training.

Treatment Response → Prognosis

Response Type	Prognosis
Rapid improvement	Excellent
Steady improvement	Good
Partial response	Variable
Minimal response	Mechanism needs re-evaluation
No response	Guarded; search for underlying cause

Section 4—Integrating Prognostic Indicators Into a Single Trajectory

This is where the magic happens.

1. High Recovery Potential If:
   - Normal or mildly abnormal baseline
   - Strong RSA
   - Good Phase II

- Present Phase IV
- PP stable
- Sudomotor near-normal
- Stress tolerance moderate
- Responds to therapy

2. Chronic Stable If:
   - Moderate abnormalities across tests
   - Timing delays
   - Mild neuropathy
   - Symptoms consistent but manageable
   - Partial response to therapy

3. Guarded If:
   - Abnormal-Abnormal across multiple domains
   - Absent overshoot
   - Global sudomotor loss
   - Severe heat or postprandial intolerance
   - Weak response to therapy
   - Progressive symptoms

Section 5—The Prognostic Matrix (The Summary Grid)

Domain	Good Prognosis	Variable Prognosis	Guarded Prognosis
Baseline	Stable	Mild instability	Chaotic
Deep breathing	Strong RSA	Moderate	Flat
Valsalva II	Strong	Moderate	Weak
Valsalva IV	Present	Delayed	Absent

Domain	Good Prognosis	Variable Prognosis	Guarded Prognosis
PP upright	Stable	Narrow	Collapse
Sudomotor	Normal	Patchy	Global failure
Heat tolerance	Mild	Moderate	Severe
Food tolerance	Stable	Moderate OI	Collapse
Stress response	Manageable	Moderate	Shutdown
Treatment response	Strong	Partial	None

Section 6—Special Considerations: Etiology Matters

1. Post-Viral Dysautonomia
   - Prognosis: good with time
   - Recovery trajectory: months to years
   - Pattern: normal-abnormal often

2. Autoimmune Dysautonomia
   - Prognosis: variable
   - Improvement depends on immune control

3. Diabetic Autonomic Neuropathy
   - Prognosis: depends on glycemic control
   - Pattern: neuropathy dominant

4. Genetic Dysautonomia
   - Prognosis: stable/chronic
   - Symptoms managed, not reversed

5. Idiopathic POTS
- Prognosis: generally favorable
- Most improve significantly with targeted therapy

Science Snapshot—Prognosis in a Single Table

Mechanism	Typical Prognosis	Reason
Preload failure	Excellent	Treatable
Hyperadrenergic	Good/Variable	Responsive to meds
Vagal impairment	Moderate	Slow nerve recovery
Baroreflex dysfunction	Variable	Timing-dependent
Small fiber neuropathy	Variable	Depends on cause
Global failure	Guarded	Multisystem

Translation—What This Chapter Really Means
- Prognosis isn't guesswork—it's mechanistic.
- Certain failure modes recover beautifully (preload failure, hyperadrenergic).
- Others improve slowly (vagal, baroreflex).
- Some depend entirely on root cause (neuropathy, autoimmune).
- The more domain involvement, the more guarded the prognosis.
- Treatment response is the strongest predictor of long-term outcome.

Tilt Tip—How to Use Prognostic Knowledge in Real Life
- Track your improvement markers, not just symptoms.
- Focus on mechanisms that can be improved: preload, tone, and vagal reserve.

- Recognize that heat and food are prognostic stressors—manage them aggressively.
- Use your treatment response as a real-time indicator of trajectory.
- Advocate for deeper evaluation if your sudomotor or baroreflex status worsens.

PART V
THE COUNTEROFFENSIVE: MANAGEMENT BY PHENOTYPE & MECHANISM

Treatment isn't a vibe—it's strategy.

CHAPTER 15
POTS: PRECISION TREATMENT BY SUBTYPE

Because "POTS" isn't one diagnosis—it's four different problems wearing the same name tag.

Introduction—The POTS Problem Is Not the Name; It's the Oversimplification

Postural Orthostatic Tachycardia Syndrome (POTS) is treated by many clinicians as though it is one single disease. This is the medical equivalent of saying all cars drive the same way because they all have wheels.

POTS is a syndrome—not a mechanism, not a cause, not a final answer.

It is a heart rate response under orthostatic conditions, driven by different physiologic failures, presenting with overlapping symptoms, and requiring entirely different treatment strategies.

Treating all POTS patients the same is like treating every fever with the same antibiotic—lazy, ineffective, and guaranteed to disappoint everyone.

This chapter breaks POTS into four dominant physiologic subtypes, explains how to identify each one from testing patterns, and—most importantly—explains how to treat each subtype without accidentally worsening the others.

Because nothing says "clinical disaster" like treating hyperadrenergic POTS with fludrocortisone, or neuropathic POTS with beta-blockers alone.

Welcome to precision POTS management.

Section 1—The Four Major POTS Subtypes

There are four dominant mechanistic phenotypes:
1. Preload failure POTS (the pooler phenotype)
2. Hyperadrenergic POTS (the adrenaline volcano)
3. Neuropathic POTS (the nerve failure variant)
4. Secondary or Comorbid POTS (the root-cause-driven type)

Each relies on completely different physiology. Treating POTS without identifying the subtype is the definition of flying blind with a stethoscope.

Section 2—Subtype 1: Preload Failure POTS

The classic, the common, the endlessly misunderstood.

This is the most prevalent subtype. These patients simply do not return enough blood to the heart when upright—period.

Mechanism:
- Excessive venous pooling
- Low stroke volume
- The heart compensates by increasing HR

Not anxiety. Not deconditioning. Physics.

Diagnostic Signature

Test	Finding
Baseline	Elevated HR, narrow PP
Deep Breathing	Normal or mildly low
Valsalva	Strong HR rise, weak BP stabilization
Standing	HR↑ ≥30–50 bpm, PP↓, BP stable
Sudomotor	May be normal or mildly reduced

This subtype is *desperate* for more blood volume and better vascular tone.

Core Treatment Strategy: Fix the Physics

1. Salt and Water Loading
   - 3–10 g sodium/day
   - 2–3+ liters fluid/day
   - Mechanism: increase plasma volume

2. Compression
   - 20–30 mmHg minimum
   - Waist-high > thigh-high > calf
   - Mechanism: reduce venous pooling

3. Volume-Expanding Drugs

Medication	Purpose
Fludrocortisone	Expands blood volume
Desmopressin (PRN)	Short-term volume boost
IV saline (if needed, selective)	Rescue therapy

4. Vasoconstrictors

Medication	Mechanism
Midodrine	Increases peripheral resistance
Droxidopa	Increases norepinephrine availability

These improve vascular tone so blood stops abandoning ship.

5. Exercise Rehab (Levine/Modified Programs)
- Recumbent→semi-recumbent→upright sequence
- Increases stroke volume over time

6. Avoid Worsening Factors
- Heat
- Large meals
- Dehydration
- Sudden standing

Not Helpful (or Harmful)

Medication	Why It's Bad Here
Beta-blockers (high dose)	Worsen low stroke volume
Clonidine	Drops BP in poolers
Aggressive vasodilation meds	Increase pooling

Precision matters.

Section 3—Subtype 2: Hyperadrenergic POTS

The adrenaline-fueled carnival of physiology.

This subtype is defined by excessive norepinephrine activity, typically due to sympathetic overactivation, impaired reuptake, mast cell activation, or compensatory mechanisms gone haywire.

Mechanism:
- Too much norepinephrine
- β-receptor overdrive
- Excessive HR and BP responses
- Tremor, heat, anxiety-like symptoms

Not psychological. Chemical.

Diagnostic Signature

Test	Finding
Baseline	Tremor, high HR, wide PP, sympathetic noise
Standing	HR↑ + BP↑ or stable, sometimes spikes
Valsalva	Exaggerated HR response
Sudomotor	High output/Hyper-conductance
Labs (when done)	NE >600 pg/mL upright

Hyperadrenergic POTS is loud, dramatic, and very responsive to targeted therapy.

Core Treatment Strategy: Turn Down the Adrenaline

1. Beta-Blockers
Best options:
- Propranolol
- Nadolol
- Atenolol

Mechanism:
- Blunt HR spikes
- Block NE effect

2. Alpha-2 Agonists

Medication	Effect
Clonidine	Reduces central sympathetic output
Guanfacine	Similar, longer acting

3. Mast Cell Stabilization (when appropriate)
- H1 blockers
- H2 blockers
- Cromolyn sodium
- Montelukast

4. Volume Optimization
- Increase plasma volume
- Prevents compensatory surges

5. Sleep & Adrenal Control
- Treat poor sleep
- Treat stress triggers
- Cortisol modulation when necessary

Not Helpful (or Harmful)

Medication	Why It's Bad
High-dose midodrine	Worsens HTN episodes
Fludrocortisone (sometimes)	May increase adrenergic drive
Stimulants	Turbocharge symptoms

Hyperadrenergic POTS requires calm, not volume overload.

Section 4—Subtype 3: Neuropathic POTS

The "wiring problem" version, where the nerves simply don't deliver the signals.

This subtype is caused by partial or patchy sympathetic denervation—particularly in the lower extremities.

Mechanism:

- Impaired vasoconstriction
- Selective sympathetic nerve loss
- Distal pooling due to denervation
- Compensatory HR increase

Essentially, the nerves don't tell the vessels to tighten.

Diagnostic Signature

Test	Finding
Baseline	Mild tachycardia or normal
Deep Breathing	Normal or mildly impaired

Test	Finding
Valsalva	Weak Phase II, delayed responses
Standing	HR↑ with PP↓; possible mild OH
Sudomotor	Distal or patchy hypofunction

Neuropathic POTS is the "quiet" subtype—less dramatic than hyperadrenergic, but more structurally grounded.

Core Treatment Strategy: Support the Failing Nerves

1. Vasoconstrictors
- Midodrine
- Droxidopa

2. Volume Expansion
- Salt
- Water
- Fludrocortisone

3. Neuropathic Support

Intervention	Mechanism
B vitamins	Nerve repair
Alpha-lipoic acid	Small fiber support
Lifestyle anti-inflammatory	Reduces nerve injury

4. Compression

Absolutely required. This is a pooling-driven subtype.

5. Targeted Exercise
Goal:
- Improve venous return
- Increase muscle pump function

Not Helpful

Medication	Why It's Bad
High-dose beta-blockers	Worsen fatigue
Clonidine	Suppresses already-weak sympathetic tone

Neuropathic POTS wants support, not suppression.

Where Mestinon Fits in Neuropathic POTS
Pyridostigmine (Mestinon) is particularly useful in neuropathic POTS because it strengthens the parasympathetic braking system and improves baroreflex coordination—two deficits that define this subtype. It does not fix the neuropathy, but it stabilizes timing, reduces desperation tachycardia, and helps the heart stop overreacting to inadequate vascular support.

In short: Mestinon gives the autonomic system enough modulation to behave less like an understaffed fire department and more like a team with an actual dispatch plan.

Section 5—Subtype 4: Secondary/Comorbid POTS
Because sometimes POTS is the symptom, not the disease.

These cases occur when another condition causes POTS physiology.

Common culprits include Ehlers–Danlos syndrome, mast cell activation disorders, autoimmune disease, post-viral or post-infectious syndromes, small fiber neuropathy, anemia, chronic inflammatory conditions, and endocrine abnormalities.

Mechanism: The underlying condition destabilizes vascular tone, connective tissue integrity, immune signaling, nerve function, and volume regulation, producing POTS physiology.

Diagnostic Signature

Feature	Clue
Multisystem involvement	Beyond orthostasis
Connective tissue findings	Hypermobility
Immune markers	Autoimmune patterns
Post-infectious timeline	Viral/Immune trigger
Gastrointestinal involvement	Motility + pooling

Secondary POTS cannot be fixed without addressing the root cause. Treating only the HR rise is like putting dryer sheets in a burning building—pleasant, but not helpful.

Section 6—Cross-Subtype Trap: Misdiagnosis & Mistreatment

Misidentifying the subtype leads to predictable disasters.

Examples of Common Clinical Mishaps

Mistake	Consequence
Giving beta-blockers to pure preload failure	Worsens fatigue & dizziness

Mistake	Consequence
Giving midodrine to hyperadrenergic POTS	Hypertensive spikes
Giving fludrocortisone to MCAS patients	Worsens fluid retention & inflammation
Ignoring neuropathy	Lifelong under-treatment
Calling everything anxiety	Zero physiologic management

Precision prevents disaster.

Section 7—How to Identify Subtype From the Test Results

Subtype Map

Test Domain	Preload Failure	Hyper-adrenergic	Neuropathic	Secondary
Baseline	High HR	High HR + noise	Variable	Depends on cause
Deep Breathing	Normal	Noisy	Normal/Mild	Variable
Valsalva	Strong HR, weak BP	Exaggerated	Weak Phase II	Mixed
Standing	HR↑ + PP↓	HR↑ + BP↑	HR↑ + mild OH	Variable
Sudomotor	Normal/Mild	High	Distal/Patchy	Depends on cause

Section 8—Treatment Matrix by Subtype
Master Table

Therapy	Preload Failure	Hyper-adrenergic	Neuropathic	Secondary
Salt/Water	★★★★★	★★☆☆☆	★★★★☆	Varies
Compression	★★★★★	★★★☆☆	★★★★★	★★★★☆
Midodrine	★★★★☆	★☆☆☆☆	★★★★★	★★★☆☆
Fludro-cortisone	★★★★★	★☆☆☆☆	★★★★☆	Varies
Beta-blockers	★★☆☆☆	★★★★★	★★☆☆☆	Varies
Clonidine/Guanfacine	☆☆☆☆☆	★★★★★	☆☆☆☆☆	Situational
MCAS therapy	★★★☆☆	★★★★☆	★★★☆☆	★★★★★
Exercise rehab	★★★★★	★★★★☆	★★★★★	★★★★☆

The right treatment is transformative. The wrong treatment is catastrophic.

Science Snapshot—POTS Subtypes in One Table

Subtype	Mechanism	Key Clue	Best Treatment Focus
Preload failure	Low SV	PP↓	Volume + tone
Hyperadrenergic	High NE	Tremor + BP↑	Blockers + calmers
Neuropathic	Denervation	Distal sweat loss	Tone + volume
Secondary	Root cause	Multisystem	Treat underlying

Translation—What This Chapter Really Means

- POTS is not a diagnosis—it's a physiologic reaction pattern.
- There are four major subtypes, each requiring different treatments.
- Mismanagement occurs when clinicians treat heart rate instead of mechanism.
- Matching treatment to subtype transforms outcomes.
- Precision POTS care is the difference between functioning and floundering.

Tilt Tip—How to Use Subtype Knowledge in Real Life

- Identify the subtype from the test results—not the symptoms.
- Do not accept one-size-fits-all treatment.
- Adjust daily strategies based on the physiology:
    1. Poolers: hydrate and compress.
    2. Hyperadrenergics: calm the chemical volcano.
    3. Neuropathics: support failing nerves.
    4. Secondary POTS patients: target the root cause.
- Track the subtype's response to therapy—this is the best measure of progress.

CHAPTER 16
MECHANISM-DIRECTED THERAPY

Because treating dysautonomia by symptoms is chaos, but treating it by mechanism is strategy.

Introduction—This Is Not "Try It and See." This Is Targeted Warfare.

Most dysautonomia treatment fails not because the patient is complex, but because clinicians throw random medications at symptoms like they're guessing on a multiple-choice exam. In the words of Sun Tzu, this is fighting a war without knowing the battlefield.

Everything in dysautonomia has a mechanism, and every mechanism has a correct intervention, numerous useless interventions, and a few catastrophically wrong interventions. Mechanism-directed therapy is how you stop guessing and start strategizing.

The rules are simple:
1. Identify the dominant failure mode.
2. Match therapy to physiology—not symptoms.
3. Treat the underlying insufficiency.
4. Never suppress a system that's already weak.
5. Never stimulate a system that's already over-firing.

Let's build the entire treatment playbook—mechanism first.

Section 1—Five Mechanistic Categories of Dysautonomia Treatment

Every intervention falls into one of five categories:
1. Volume expansion (fixes preload failure)
2. Vascular tone support (fixes sympathetic under-activation)
3. Adrenergic modulation (fixes sympathetic excess)
4. Parasympathetic enhancement (improves vagal braking)
5. Neural support and repair (supports small fiber + autonomic nerves)

Most patients need a blend. But each category must be used with intention, not desperation.

Central vs. Peripheral Treatment Strategies: Signal Control vs. Structural Support

Some therapies alter central autonomic signaling—how forcefully the system fires, amplifies, or suppresses responses (e.g., carvedilol, tricyclic antidepressants). Others act peripherally, improving preload, vascular tone, and end-organ perfusion without touching central drive. Confusing these is one of the fastest ways to worsen dysautonomia.

Lowering heart rate or calming symptoms does not guarantee improved circulation. If the problem is flow, treat peripherally first. If the problem is signal, consider central modulation. When uncertain, assume tachycardia is compensatory until proven otherwise.

Section 2—Mechanism 1: Preload Failure (Low Stroke Volume)

The Pooler Phenotype: the most common and the most fixable.

Problem: Not enough blood returns to the heart. Gravity wins. The patient loses.

Mechanism:
- Venous pooling
- Low plasma volume
- Reduced stroke volume
- Compensatory tachycardia

Targeted Interventions:
1. Salt Loading
3–10 g sodium/day
Raises plasma volume.

2. Hydration
2–3+ liters fluid/day
Boosts intravascular volume.

3. Fludrocortisone
Expands plasma volume by enhancing sodium retention.

4. Desmopressin (Intermittent)
Short-term volume rescue for bad days.

5. Compression Therapy
- Waist-high > thigh-high > calf
- 20–30 mmHg minimum

Reduces venous pooling and instantly increases preload.

6. Exercise Reconditioning
Start horizontal → semi-recumbent → upright
Goal: increase stroke volume over time.

7. Midodrine
If vascular tone is also impaired.

Do not use high-dose beta-blockers, clonidine, or vasodilators; these worsen low stroke volume.

Section 3—Mechanism 2: Sympathetic Underactivation
The "my nerves forgot to vasoconstrict" problem.

Problem: Sympathetic fibers cannot constrict blood vessels.

Mechanism:
- Impaired norepinephrine release
- Weak vascular tone
- Early BP drops
- Poor orthostatic tolerance

Targeted Interventions:
1. Midodrine
Direct vasoconstriction.

2. Droxidopa
Boosts norepinephrine availability in postganglionic synapses.

3. Volume Expansion
Salt + water + fludrocortisone.

4. Physical Counter-Maneuvers
- Calf pumps
- Isometric contractions
- Leg crossing

All increase venous return.

5. Compression Garments
Critical for these patients.

Do not use clonidine, high-dose beta-blockers, or central sympatholytics; these suppress an already weak system.

Section 4—Mechanism 3: Sympathetic Excess (Hyperadrenergic States)
The adrenaline volcano: loud, chaotic, sometimes misunderstood.

Problem: Too much norepinephrine firing.

Mechanism:
- Impaired reuptake
- Overactive sympathetic signaling
- Adrenal overresponse
- *Compensatory hyperadrenergic surges from preload stress*

Targeted Interventions:
1. Beta-Blockers
Low-dose propranolol or nadolol.

2. Alpha-2 Agonists
- Clonidine
- Guanfacine

Reduce acute central sympathetic output.
- Methyldopa

Reduces chronic central sympathetic output. Use only in confirmed hyperadrenergic physiology with preserved blood pressure. Never first-line; may worsen neuropathic, low-volume, or venous-pooling states by suppressing compensation. Slower and heavier than clonidine or guanfacine—appropriate only when true braking is needed.

3. Selective Chronotropic Control
- *Corlanor (ivabradine)*

Lowers sinus-node firing without suppressing sympathetic tone, vascular reflexes, or baroreflex signaling. Appropriate when tachycardia is limiting but blood pressure and perfusion are preserved.

4. Mast Cell Stabilization (when relevant)
- H1 blockers
- H2 blockers
- Cromolyn
- Montelukast

Because MCAS can fuel NE release.

5. Volume Optimization
Moderate salt + fluid.
Hyperadrenergics often compensate for preload issues.

6. Sleep Optimization
Because nothing spikes NE like poor REM sleep.

Do not use midodrine (may induce hypertensive spikes), high-dose fludrocortisone (can worsen surges), or stimulants.

Hyperadrenergic physiology wants calming, not cranking.

Section 5—Mechanism 4: Parasympathetic Impairment

The system has an accelerator but no brakes.

Problem: The vagus nerve cannot apply adequate braking force.

Mechanism:
- Reduced HR variability
- Poor respiratory sinus arrhythmia
- Weak Valsalva HR brake
- Slow recovery from stressors

Targeted Interventions:

1. Vagal Training
   - Paced breathing (5–6 breaths/min)
   - Resonance frequency training
   - Biofeedback

2. Anti-inflammatory Diet & Lifestyle
Because vagal tone decreases when systemic inflammation rises.

3. Gastrointestinal Optimization
Vagus connects gut → brain.
Improve gut motility → improve vagal signaling.

4. Physical Conditioning
Regular aerobic activity increases vagal tone.

5. Sleep Enhancement
Deep sleep boosts parasympathetic dominance.

Do not use medications that blunt vagal activity (e.g., certain anticholinergics), high reliance on acute stimulants, or adrenaline-spiking habits.

Parasympathetic enhancement requires consistency, not intensity.

Section 6—Mechanism 5: Small Fiber Neuropathy

The wiring problem: when the communication lines are damaged.

Problem: Small unmyelinated fibers fail to transmit autonomic signals.

Mechanism:
- Impaired vasoconstriction
- Abnormal sweating
- Poor vascular tone
- Sensory symptoms
- Patchy or distal autonomic failure

Targeted Interventions:
1. Underlying Etiology Management
Autoimmune? Inflammatory? Metabolic? → Fixing the root cause produces the best improvement.

2. Nerve Repair Support

Supplement/Intervention	Mechanism
Alpha-lipoic acid	Antioxidant + nerve repair
Acetyl-L-carnitine	Mitochondrial support
B-complex	Nerve metabolism
Omega-3s	Anti-inflammatory

3. Vascular Tone Support
Midodrine or droxidopa.

4. Physical Conditioning
Improves muscle pump, mitigates denervation consequences.

5. Compression Therapy
Critical for distal pooling.

Do not use high-dose sympatholytics; these suppress a system that's already struggling to fire.

Section 7—Mechanism 6: Baroreflex Timing Dysfunction
The reflex works—but too late, too early, or unpredictably.

Problem: The baroreflex is mistimed.
Mechanism:
- Unstable blood pressure
- Oscillation
- Delayed responses
- Excessive or insufficient corrections

Targeted Interventions:
1. BP Stabilization
    - Midodrine (for undercorrection)
    - Clonidine or guanfacine (for overcorrection)

2. Vagal Training
Improves baroreflex sensitivity.

3. Controlled Breathing Practices
Reduces oscillation.

4. Avoid Triggers
- Stress
- Heat
- Sudden postural changes

Do not use rapid-onset vasoactive drugs. They exacerbate timing mismatches.

Section 8—Paradoxic Parasympathetic Syndrome (PPS)

PPS management prevents the baroreflex from reaching the delayed-detection threshold that triggers the vagal overshoot.

Strategic Goals:
1. Stabilize venous return.
2. Avoid prolonged sympathetic strain.
3. Interrupt pre-collapse phases early.

Targeted Approaches:
- Salt/Fluid loading
- Compression garments
- Counterpressure maneuvers
- Avoid static standing
- Pre-hydration before heat or exertion
- Pharmacologic support: fludrocortisone, midodrine, pyridostigmine, low-dose beta blockade depending on phenotype

The key to preventing PPS is preventing the baroreflex from panicking.

Section 9—Cross-Mechanism Failures: Combination Therapy Strategy

When two mechanisms fail at once, treatment must be prioritized based on the dominant driver.

If preload failure + hyperadrenergic
- Fix volume first
- Only then modulate NE

If neuropathy + preload failure
- Volume support
- Tone support
- Nerve repair

If vagal impairment + hyperadrenergic
- Calm sympathetic first
- Strengthen vagal reserve second

Combination therapy succeeds only when you treat the initiating mechanism first—not the loudest symptom.

Section 10—Mechanistic Treatment Matrix

Mechanism	Best Interventions	Avoid These
Preload failure	Salt, fluids, fludrocortisone, compression, midodrine	High-dose beta-blockers, clonidine
Sympathetic underactivation	Midodrine, droxidopa, compression	Alpha-2 agonists, heavy beta-blockade
Sympathetic excess	Beta-blockers, clonidine, mast-cell therapy	Midodrine, high-salt diets (sometimes)

Mechanism	Best Interventions	Avoid These
Vagal impairment	Breathing training, gut optimization, sleep	Anticholinergics, stimulants
Small fiber neuropathy	Nerve repair, tone support, volume	Sympatholytics
Baroreflex dysfunction	Controlled tone modulation	Rapid-acting vasoactives

This is the clinical "cheat sheet" everyone should be using but isn't.

Section 11—The Sequence Rule: What to Treat First

To avoid worsening the wrong system, follow this order:

1. Treat preload failure first—without adequate stroke volume, nothing else matters.

2. Stabilize tone second—fix underactivation or excess sympathetic activity.

3. Strengthen vagal control third—parasympathetic support works only when the basics are fixed.

4. Repair nerve function over time—neuropathy takes longer but is essential for long-term stability.

5. Fine-tune the baroreflex last—this improves once volume and tone normalize.

Precision sequencing prevents clinical disasters.

Science Snapshot—Mechanism → Therapy

Mechanism	Core Problem	Best Therapy
Preload failure	Low SV	Salt, fluid, compression
Sympathetic underactivation	Weak vasoconstriction	Midodrine, droxidopa
Sympathetic excess	NE overload	β-blockers, clonidine
Vagal impairment	Poor braking	Vagal training
Small fiber neuropathy	Nerve loss	Repair + tone
Baroreflex dysfunction	Timing errors	Stabilization

Translation—What This Chapter Really Means

- Dysautonomia treatment works when it is mechanism-directed, not symptom-directed.
- Preload failure is fixed with volume and compression.
- Sympathetic underactivation gets tone support, not suppression.
- Hyperadrenergic physiology needs calming, not compression alone.
- Vagal impairment requires training, not drugs.
- Neuropathy requires repair, not brute-force pharmacology.
- Baroreflex issues need stabilization, not guesswork.
- The order of treatment matters just as much as the treatment itself.

Tilt Tip—How to Use Mechanism-Directed Therapy in Real Life

- Know your dominant mechanism before accepting treatment.
- Use interventions that support your physiology, not ones that fight it.

- Track responses to see if you're treating the right mechanism.
- Adjust therapy with intention, not desperation.
- Never accept "POTS treatment" without subtype identification.
- Bring this chapter to your clinician if necessary—they can thank you later.

CHAPTER 17
ADVANCED MANAGEMENT STRATEGIES

Because once you know the mechanisms, the real game is how you stack, layer, and sequence them without crashing.

Introduction—You've Graduated from "What Is This?" to "How Do I Control It?"

Basic dysautonomia management is to drink more water, eat more salt, wear compression garments, and try not to faint in public.

Advanced management is understanding which days to push, which days to coast, how to layer therapies without tripping over side effects, how to build your life around physiology instead of constantly losing to it, and how to adjust strategy when your autonomic system decides to change the rules mid-game.

This chapter assumes you've identified your dominant failure mode, you understand your triggers, and you've got basic treatment in place.

Now we move to tactical planning, multi-mechanism management, crash prevention, and long-term optimization.

This is the "how to live with this like a strategist, not someone being held prisoner by their autonomic nervous system" chapter.

Section 1—The Management Pyramid: Stop Doing Level 3 Before Level 1

Think of dysautonomia management as a pyramid:
1. Foundation: physiology basics
2. Middle: mechanism-directed Interventions
3. Top: advanced modifiers and fine-tuning

Most people start at the top (beta-blocker, midodrine, random supplement stack) and skip the foundation. That rarely ends well.

The Dysautonomia Management Pyramid

Level	Focus	Examples
3	Fine-tuning & advanced therapies	Meds combos, MCAS protocols, neurorehab
2	Mechanism-directed treatment	Volume, tone, adrenergic modulation
1	Basic physiology & environment	Hydration, salt, pacing, sleep, temperature, posture

If Level 1 is unstable, Levels 2 and 3 will wobble no matter how "correct" they are.

Section 2—Level 1: Advanced Foundation Work (Not Actually Basic)

You already know the basics—salt, water, compression, sleep, and pacing. Advanced management means getting granular and deliberate about them.

1. Timing, Not Just Volume
It's not just how much you drink or salt—it's when.

Time	Strategy	Why It Matters
Morning	Front-load fluids + salt	Preload for the worst orthostatic period
Pre-upright activity	500–750 mL with sodium	Stabilizes before stressor
Pre-meal	Fluids without overdoing	Supports postprandial perfusion
Evening	Moderate intake	Avoid nocturnal overload, retain sleep

2. Temperature Zoning
You are not "sensitive." You are a heat-intolerant, vasodilatory physics experiment.

Build your life as if this is true—because it is: keep the bedroom cool, use a cooling vest or neck wraps for outings, place fans in key locations (desk, living room, bathroom), and avoid direct sun combined with static standing—the worst combination.

3. Upright Budgeting
You have a limited daily upright budget. Spend it like money, not like oxygen. Standing is expensive. Sitting costs less. Reclining is recovery.

Section 3—Level 2: Mechanism-Directed Strategy, But Smarter

You've already matched treatment to preload failure, sympathetic under/overactivity, vagal impairment, neuropathy, and baroreflex issues.

Advanced management = combining these without torpedoing yourself.

Principle 1—One Lever at a Time
Never change your dose, medication, compression, and exercise level in the same week and then try to figure out why you crashed.

One new variable at a time. Everything else stays steady.

Principle 2—Sequence Before Dose
For example:
- Preload failure + hyperadrenergic: fix preload first, then gently modulate adrenergic excess; do NOT start with heavy beta-blockade on a dry patient.
- Neuropathic + baroreflex delay: improve tone + preload, then gently stabilize baroreflex behavior.

Principle 3—Respect Recovery Windows
Autonomic systems adapt over weeks to months, not hours to days. If you don't give physiology time to respond, it cannot show you what works.

Section 4—Level 3: Advanced Modifiers (For When the Basics Are Already Locked In)

These are not first-line treatments. They are refinement tools.

1. Precision Beta-Blocker Use
Instead of, "You have POTS—here's a beta-blocker forever," use targeted dosing, symptom-based timing, and subtype-specific medication choices.

Scenario	Approach
Hyperadrenergic mornings	Low-dose beta-blocker AM only
Exercise-triggered tachycardia	Use pre-activity
Preload failure POTS	Avoid high-dose; only micro-dose as adjunct

2. Selective Vasoconstriction

Midodrine or droxidopa should be timed to precede upright demand—not taken randomly. Use them before workdays, shopping, events, or appointments, not before lying on the couch.

3. Tiered Exercise Progression

Instead of, "Do cardio," use a three-stage progression.

Stage	Position	Goal
1	Supine/Recumbent	Tolerance without crash
2	Semi-recumbent/Seated	Build endurance
3	Upright	Real-world translation

The goal is *more blood moved with less autonomic chaos*, not "more steps."

Section 5—Trigger Engineering: Deliberate Exposure Without Meltdown

Advanced management is not avoidance-only; it's controlled exposure.

1. Heat Strategy:
- Pre-hydrate + sodium
- Use compression
- Limit exposure duration
- Schedule recovery time afterward

2. Postprandial Strategy:
- Split meals into smaller portions
- Lower carb load per meal
- Avoid heavy upright tasks right after food
- Evaluate whether certain macronutrients wreck you worse

3. Stress Strategy:
- Treat stress like a physiologic dose, not a moral failure
- Schedule cognitively heavy tasks earlier in the day
- Avoid combining: stress + heat + standing + fasting

4. Standing Strategy:
- Pre-hydrate
- Brace muscles (calf/quad squeeze)
- Subtle shifts in position
- Use a stool/lean when possible (leaning ≠ weakness; it's a preload strategy)

Section 6—Multi-Mechanism Patients: How Not to Lose Your Mind

Most real people have preload failure, some degree of hyperadrenergic behavior, maybe mild neuropathy, and occasional baroreflex weirdness.

You can't treat everything at once. You prioritize.

Prioritization Framework
1. Can they stand at all? → Preload and tone first.
2. Are they constantly in "fight or flight"? → Adrenergic modulation next.
3. Do they have severe HR/BP unpredictability despite #1 and #2? → Baroreflex targeting.

Do sensory symptoms and sudomotor loss dominate? → Neuropathy support integrated along the way.

Section 7—Crash Management: What to Do When the System Mutinies

Crashes are inevitable. Advanced management means you shorten them, you reduce their severity, and you learn from them.

Crash Triage Framework

Question	Action
Was there a clear trigger?	Adjust that domain going forward
Is this worse than usual?	Consider infection, flare, new med, sleep loss
Are vitals unstable at rest?	Reassess meds & volume urgently
Is this a pattern?	Modify baseline strategy, not just rescue

Crash Rescue Tools (Mechanism-Based):
- Preload failure crash → fluids + salt + compression + recumbency
- Hyperadrenergic crash → calm sensory load, maybe beta-blocker or clonidine (per plan)
- Heat crash → aggressive cooling + horizontal
- Postprandial crash → horizontal, time, fluids, avoid stacking events afterward

Rescue plans should be pre-written, not invented mid-crash.

Section 8—Longitudinal Tracking: Treat It Like a Research Project

Advanced management is impossible without data. Not obsessively, but systematically.

What to Track

Domain	Examples
Orthostatic symptoms	Time to dizziness, severity
Triggers	Heat, meals, stress, cycle
Daily function	Hours upright, activity blocks
Treatment	Dose, timing, changes
Crashes	Frequency, triggers, duration

This is not "overthinking it." This is generating a longitudinal autonomic dataset.

Section 9—Team Strategy: When and How to Involve Other Specialists

Dysautonomia often sits at the intersection of gastroenterology, rheumatology, cardiology, neurology, psychiatry, and endocrinology. Advanced management means knowing what you want from each specialty.

Example: Ask Strategically
- Neurology → nerve function, small fiber evaluation, central causes
- Cardiology → rhythm, structure, safe exercise limits
- Rheumatology/Immunology → autoimmune contributors
- GI → motility and absorption issues impacting nutrition + meds

You're not there to hand them your identity and hope for magic. You're there to extract specific, mechanism-relevant information.

Science Snapshot—Advanced Management in One Table

Problem	Mechanism	Advanced Strategy
Daily unpredictable crashes	Baroreflex + load stacking	Adjust triggers + timing, not just meds
Morning wipeout	Overnight hypovolemia	Front-load fluids/salt, adjust sleep & meds
Post-meal collapses	Splanchnic pooling	Smaller, lower-carb meals, upright budgeting
Heat-triggered meltdown	Vasodilation + poor cooling	Environmental control + pre-hydration
Flare with stress	Hyperadrenergic physiology	Adrenergic modulation + pacing of cognitive load

Translation—What This Chapter Really Means

- What looks like instability is physiology operating on a tight, unforgiving budget.
- Advanced management is not "more meds, more compression, more everything."
- The goal is the right lever, at the right dose, with the right timing—while minimizing collateral damage.
- Foundation work (hydration, salt, temperature, pacing, sleep) is not optional.
- Complex patterns untangle once you prioritize mechanisms and sequence interventions.
- Crashes become shorter and less terrifying when you have a scripted response.
- Data beats vibes. Strategy beats improvisation.

Tilt Tip—How to Use Advanced Strategy Without Burning Out

- Treat your body like a high-performance system with narrow margins, not a broken machine.
- Build routines that protect your physiology so you don't have to constantly firefight.
- Make one strategic change at a time and give it enough time to show what it does.
- View every flare as a lab report: what combination of triggers, load, and context produced it?
- Remember: this is long-term management, not a 2-week challenge. Consistency is the intervention.

CHAPTER 18
CLINICAL CASE INTEGRATION

Because knowing each mechanism is fun, but watching them collide inside a real human is where the plot actually gets good.

Introduction—Welcome to the Autonomic Escape Room

You now know how the autonomic system works, how it fails, how we test it, how we interpret those tests, and how we target the mechanisms.

Congratulations. You've been assembling puzzle pieces. Now we zoom out and look at the whole picture. This is where dysautonomia stops being something you've read about—and starts being something you actually understand.

This chapter teaches you how to synthesize symptoms, triggers, resting state physiology, deep breathing, Valsalva, standing/tilt, sudomotor, medications, and real-world conditions into a single dominant mechanistic narrative.

This is not guesswork. This is physiology behaving exactly as expected—if you know how to read it.

Section 1—Why Integration Matters More Than Any Single Test

Single tests whisper. Integrated physiology confesses (loudly).

Isolated interpretation fails because symptoms can be misleading, triggers overlap, different mechanisms produce similar sensations, medications distort signals, and some abnormalities compensate for others.

But when everything is combined, a consistent mechanism emerges.

Section 2—The Autonomic Integration Framework
Every real-world case must be interpreted in six steps:
1. Symptoms → suggestive but never diagnostic
2. Triggers → reveal instability domains
3. Resting state → baseline truth serum
4. Deep Breathing → parasympathetic reserve
5. Valsalva → sympathetic strength + baroreflex timing
6. Standing/Tilt → real-world physiological collapse patterns
7. Sudomotor → structural vs. functional involvement
8. Medication Context → what's bending the curve
9. Integration → reveal the dominant failure mode

Let's build integrated patterns you can apply to any case—without violating HIPAA or offering personal anecdotes.

Section 3—Archetype 1: The Pooler (Preload Failure Dominance)
A physics problem masquerading as a mystery.

Symptoms:
- Dizziness on standing
- Tachycardia
- Fatigue, especially in the morning
- Heat intolerance
- Post-meal wobbliness

Triggers:

- Standing
- Heat
- Long conversations
- Hot showers

Testing Pattern

Test	Finding
Baseline	High HR, narrow PP
Deep Breathing	Normal or mildly low
Valsalva	Strong HR, weak BP recovery
Standing	HR↑ ≥30–50 bpm + PP↓
Sudomotor	Normal/Mild distal issues

Interpretation: Blood volume doesn't reach the heart. Stroke volume drops. HR compensates.

Management Focus:

- Fluids + salt
- Compression
- Midodrine if tone also low
- Recumbent → upright exercise

This case is the poster child of treatable dysautonomia.

Section 4—Archetype 2: The Adrenaline Volcano (Hyperadrenergic Dominance)

If resting physiology sounds like a humming transformer, this is why.

Symptoms:
- Tremor
- Heat surges
- Chest pressure
- Adrenaline "spikes"
- Jittery orthostasis

Triggers:
- Stress
- Standing
- Prolonged cognitive load
- Skipping meals

Testing Pattern

Test	Finding
Baseline	High HR, wide PP, sympathetic noise
Deep Breathing	Noisy with low amplitude
Valsalva	Exaggerated responses
Standing	HR↑ + BP↑ or stable but chaotic
Sudomotor	Hyper-conductance possible

Interpretation: Excess norepinephrine release or impaired reuptake.

Management Focus:
- Beta-blockers (low dose)
- Clonidine/Guanfacine

- Moderate volume optimization
- Anti-inflammatory + MCAS stabilization (when relevant)
- Sleep reconstruction

This case requires calming the system, not filling it.

Section 5—Archetype 3: The Quiet Denervation (Neuropathic Dominance)

The nerves simply do not fire when they're supposed to.

Symptoms:
- Orthostatic fatigue
- Cold extremities
- Sensory symptoms
- Slow recovery from standing

Triggers:
- Upright demand
- Temperature shifts
- Long days
- Fasting

Testing Pattern

Test	Finding
Baseline	Mild tachycardia or normal
Deep Breathing	Mild vagal blunting
Valsalva	Weak Phase II, delayed transitions
Standing	HR↑ + mild OH
Sudomotor	Distal loss or patchy deficits

Interpretation: Sympathetic vasoconstriction doesn't happen properly.

Management Focus:
- Midodrine/Droxidopa
- Compression
- Nerve repair support (ALA, B-complex, anti-inflammatory)
- Volume expansion
- Recumbent conditioning

These cases improve but require long-term consistency.

Section 6—Archetype 4: The Timing Error (Baroreflex Dysfunction)

A reflex that works—just not on time.

Symptoms:
- Sudden BP dips
- Oscillations
- Inconsistent day-to-day function
- "I feel fine until I don't" patterns

Triggers:
- Stress
- Positional changes
- Sudden exertion

Testing Pattern

Test	Finding
Baseline	Labile BP
Deep Breathing	Delayed trough
Valsalva	Phase transitions late or absent
Standing	BP oscillation or delayed hypotension
Sudomotor	Variable

Interpretation: The baroreflex fires—but late or unpredictably.

Management Focus:
- Stabilize tone (midodrine or clonidine, depending on direction)
- Consistent breathing work
- Avoid abrupt positional changes
- Micro-dosing strategies for vasomodulators

This case is about synchronization, not pure weakness.

Section 7—Archetype 5: Mixed Mechanisms (The Realistic Majority)

Because your body doesn't read the textbook.

Symptoms: All of the above, depending on the day.

Triggers: Multiple overlapping domains
- Heat
- Meals
- Stress

- Exertion
- Prolonged upright time

Testing Pattern: Mixed

Domain	Pattern
Baseline	Mildly abnormal HR or BP
Deep breathing	Mild parasympathetic blunting
Valsalva	Partial deficits
Standing	HR↑ + PP↓ ± BP instability
Sudomotor	Mild-to-moderate deficits

Interpretation: Most patients do not fit neatly into one box.

Mixed patterns require prioritization:
1. Treat preload failure first
2. Then stabilize tone
3. Then modulate NE
4. Then improve vagal function
5. Then target nerve repair
6. Use baroreflex strategies last

This sequence prevents overcorrection and undertreatment.

Section 8—Clinical Integration Scenarios (Hypothetical, Non-Personalized)

To show how integration works, here are non-patient-specific composite scenarios.

Scenario A—The Hot-Room Collapse
- Baseline narrow PP
- Strong Valsalva HR
- Standing HR↑40 with PP↓
- Normal sudomotor → preload failure dominates

Strategy: volume + compression + midodrine PRN + horizontal exercise

Scenario B—The Stress Meltdown
- Baseline sympathetic noise
- Valsalva spikes
- Standing HR↑ + BP↑
- Sudomotor hyper → hyperadrenergic dominance

Strategy: beta-blocker microdose + clonidine at specific times

Scenario C—The Late-Day Faceplant
- Normal morning
- Worsening evening orthostasis
- Distal sudomotor deficits → neuropathic + volume fatigue

Strategy: volume timing + tone support + nerve repair

Scenario D—The Unpredictable Yo-Yo
- Baseline BP variability
- Late-phase Valsalva errors
- Inconsistent standing response → baroreflex timing disorder

Strategy: microdosed tone modulators + breathing + slow positional transitions

These are templates, not diagnoses. The point is the pattern recognition—not the specifics.

Section 9—Medication Integration: The "Don't Cancel Your Own Treatment" Rules

You cannot stimulate the sympathetic system, suppress it, expand blood volume, and dry the patient out—simultaneously. You can try, but physiology does not reward contradictory strategies.

Medication Integration Table

Mechanism	Helpful	Harmful
Preload failure	Fluids, salt, fludrocortisone, midodrine	High-dose beta-blockers
Hyperadrenergic	Low-dose beta-blocker, clonidine	Midodrine (typically)
Neuropathic	Midodrine, droxidopa, fludrocortisone	Suppressive adrenergics
Baroreflex	Careful clonidine or midodrine	Drastic shifts
Vagal failure	Breathing, conditioning	Anticholinergics

The wrong combination is worse than no treatment at all.

Section 10—Real-World Integration: How to "Read" a Patient in 5 Minutes

1. How do they respond to standing?
    - This reveals preload vs. tone vs. adrenergic vs. timing.

Clinical Case Integration

2. How do they respond to heat?
   - Heat intolerance = poor vasoconstriction or preload failure.

3. How do they respond to meals?
   - Postprandial worsening = splanchnic pooling.

4. How is their morning vs. evening pattern?
   - Mornings worse → preload
   - Evenings worse → neuropathic fatigue or baroreflex drift

5. Do they have sensory or sudomotor abnormalities?
   - This clues in on small fiber involvement.

This is clinical integration in real time.

Science Snapshot—Integration Rules in One Table

Domain	Meaning	Mechanistic Implication
Baseline	Reserve	Sympathetic/vagal tone
Deep breathing	Brake strength	Vagal integrity
Valsalva	Sympathetic + baroreflex	Timing + force
Standing	Real-world output	Preload/Tone
Sudomotor	Structural vs. Functional	Neuropathy
Triggers	Weak points	Phenotype
Meds	Distortions	Interpretation adjustments

Translation—What This Chapter Really Means

- Autonomic physiology is complex, but predictable when viewed as a unified system.

- No single symptom or test reveals the mechanism—but the combination always does.
- Most patients have multiple mechanisms; integration identifies which one leads.
- Clinical strategy is 90% pattern recognition and 10% pharmacology.
- The goal is clarity, not perfection: identify the dominant failure mode and correct it.

Tilt Tip—How to Use Integration Without Becoming Overwhelmed

- Don't treat noise—treat the dominant mechanism.
- Don't chase daily fluctuations—they're expected.
- Look for consistent patterns across test domains.
- Use integrated interpretation to prevent over-treatment.
- Remember: physiology is not random; it's reacting exactly as designed.

PART VI
REAL-WORLD SURVIVAL: WINNING IN HOSTILE CONDITIONS

Because dysautonomia doesn't care about your schedule.

CHAPTER 19
ENVIRONMENTAL WARFARE: SURVIVING HEAT, FOOD, TRAVEL & GRAVITY'S FRIENDS

Because environmental physiology is relentless, and clinicians need to understand why their patients are ambushed by daily life.

Introduction—Our Planet Is Not a Neutral Actor

Clinicians routinely underestimate the impact of environmental conditions on autonomic instability.

Patients are often labeled as "intolerant," "sensitive," or "deconditioned" when, in reality, they are navigating a constant stream of physiologic stressors that exploit the weakest links in their autonomic architecture.

Environmental triggers—heat, meals, humidity, altitude, sensory load, travel, and gravitational demand—are not minor annoyances.

They are predictable physiologic stressors that magnify preload failure, destabilize sympathetic balance, impair baroreflex compensation, and overwhelm vagal buffering.

This chapter outlines the environmental contexts that destabilize autonomic physiology, the mechanistic reasons behind them, and the clinical strategies that improve function and safety.

Section 1—Heat: The Most Potent Environmental Aggravator

Heat is consistently the most destabilizing environmental factor for patients with any phenotype of autonomic dysfunction.

Heat triggers:
- Peripheral vasodilation, enlarging venous capacitance and reducing venous return.
- Increased sweating demand, taxing sympathetic output.
- Reduced stroke volume, requiring compensatory tachycardia.
- Cerebral perfusion reduction, worsening cognitive symptoms.
- Thermoregulatory strain, impairing autonomic responsiveness further in neuropathic or low-reserve physiology.

Heat exposure is therefore a direct physiologic provocation, not an intolerance or preference.

Clinical Implications: Patients presenting with heat sensitivity are not exaggerating; they are demonstrating a normal autonomic response occurring on an abnormal baseline.

Clinicians should recognize heat intolerance as a sign of impaired vasoconstriction, reduced effective circulating volume, sympathetic over-recruitment, and potential small fiber involvement.

Clinical Strategies:
- Encourage environmental cooling (vests, fans, AC) as a physiologic intervention, not a comfort measure.
- Reinforce adequate hydration and sodium intake prior to anticipated heat exposure.

- Recommend compression garments for heat-associated pooling.
- Advise pre-cooling for unavoidable outdoor activity.

Section 2—Food: The Splanchnic Redistribution Trap

Large meals, high-carbohydrate loads, and hot foods predictably destabilize autonomic homeostasis.

Mechanisms:

- Up to 35% of blood volume shifts to the splanchnic circulation after meals.
- Insulin-mediated vasodilation amplifies pooling.
- Hot meals add thermal vasodilation to the physiologic burden.
- Impaired autonomic compensation prolongs postprandial instability.

Clinical Implications: Patients reporting "post-meal dizziness," "post-meal tachycardia," or "afternoon collapses" are experiencing reduced effective circulating volume compounded by splanchnic loading.

Clinical Strategies:

- Counsel on smaller, more frequent meals—not as diet advice, but as hemodynamic modulation.
- Recommend balanced macronutrients to avoid excessive vasodilation.
- Encourage hydration before meals.
- Consider agents that reduce splanchnic pooling (e.g., pyridostigmine in selected phenotypes).

Section 3—Travel: The Multi-System Autonomic Stress Test

Travel combines multiple autonomic challenges: prolonged upright time, heat exposure, stress, dehydration, altitude effects, sensory load, and disrupted meals.

Mechanisms:
- Extended sitting reduces venous return.
- Cabin pressure changes alter sympathetic demand.
- Heat zones in airports increase vasodilation.
- Travel stress elevates NE release.

Clinical Implications: Travel intolerance is not psychological; it reflects the intersection of physiologic vulnerabilities.

Clinical Strategies:
- Recommend pre-travel preload optimization.
- Encourage use of compression during flights or long car rides.
- Advise environmental cooling tools for airports/hot climates.
- Reinforce structured pacing after arrival.
- Recognize the legitimacy of travel-related functional decline for disability or accommodation letters.

Section 4—Sensory Load: The Overlooked Autonomic Stress Multiplier

Bright lights, noise, crowds, and chaotic environments drive sympathetic activation.

Mechanistic Considerations:
- Sensory overload increases locus coeruleus activity → heightened NE release.
- Sympathetic surge reduces baroreflex sensitivity and widens HR variability in destabilizing patterns.
- Orthostatic vulnerability worsens when sensory processing consumes sympathetic bandwidth.

Clinical Implications: Patients describing "crowded places make me dizzy" are demonstrating sympathetic bandwidth exhaustion, not anxiety.

Clinical Strategies:
- Recommending sensory-modulated environments is a *medical intervention*, not avoidance behavior.
- Encourage periodic removal from overstimulating settings.
- Counsel on behavioral pacing strategies for high-load environments.

Section 5—Gravity's Friends: Standing, Walking, and Motion Demand

The simplest real-world actions—standing, walking slowly, waiting in line—are profound physiologic challenges for autonomic-impaired patients.

Mechanisms:
- 500–1000 mL of venous pooling occurs immediately upon standing.
- Stroke volume decreases sharply in preload failure phenotypes.
- Static standing eliminates muscle pump augmentation.
- Motion sensitivity in vestibulo-autonomic pathways amplifies instability.

Clinical Implications: Standing intolerance should be interpreted as impaired vasoconstriction, inadequate preload, neuropathic autonomic contribution, or baroreflex lag—not deconditioning.

Clinical Strategies:
- Address upright intolerance through volume restoration, tone support, and mechanistic rehab progression.
- Encourage occupational therapy consultation for task modification.

Section 6—Humidity and Altitude: Atmospheric Stressors with Predictable Physiologic Impact

Humidity impairs evaporative cooling; altitude reduces oxygen saturation and increases sympathetic demand.

Mechanisms:
1. Humidity increases thermal burden, drives vasodilation, and overwhelms compensatory mechanisms.
2. Altitude increases HR and ventilation, reduces vagal tone, and can destabilize baroreflex function.

Clinical Guidance: For patients with severe autonomic instability, high humidity or altitude travel may require pre-planning, medication adjustments, and explicit counseling about risk.

Section 7—Environmental Load Mapping

Environmental triggers rarely occur in isolation. Clinicians should conceptualize environmental stressors as cumulative hemodynamic burdens.

High-Risk Combinations

Trigger Combination	Physiologic Consequence
Heat + standing	Compounded vasodilation and pooling
Food + upright activity	Splanchnic shift + reduced preload
Heat + travel	Dehydration + tone exhaustion
Stress + heat	Sympathetic overdrive
Humidity + walking	Impaired cooling + venous pooling

Understanding cumulative load improves treatment specificity and patient counseling.

Science Snapshot — Environmental Stressors as Physiologic Load

Trigger	Mechanism	Autonomic Consequence
Heat	Vasodilation	Tachycardia, hypotension, fatigue
Food	Splanchnic pooling	Dizziness, fog
Travel	Preload reduction	Orthostatic symptoms
Humidity	Impaired cooling	Overheating, vasodilation
Altitude	Sympathetic demand	Tachycardia, instability
Sensory load	NE surge	Loss of compensatory bandwidth

Translation—What This Chapter Really Means

Environmental triggers are not "preferences" or "sensitivities"—they are predictable physiologic challenges that interact with underlying autonomic constraints. By understanding the mechanisms, clinicians can better anticipate instability and create individualized strategies that reduce crashes.

Tilt Tip—How to Use This Knowledge

Patients benefit when clinicians validate environmental physiology, provide proactive strategies rather than reactive reassurance, and integrate environmental risk into treatment plans and rehab pacing.

A stable autonomic system is created by reducing preventable physiologic stress—not by encouraging patients to "push through" hostile environments.

CHAPTER 20
REHAB WITHOUT COLLAPSE: REBUILDING AUTONOMIC STABILITY WITHOUT TRIGGERING PHYSIOLOGIC MUTINY

Because telling dysautonomia patients to "just exercise" is the clinical equivalent of asking a drowning person to "swim harder."

Introduction—Why Traditional Exercise Fails Patients with Autonomic Dysfunction

Clinicians frequently underestimate how profoundly autonomic physiology determines exercise tolerance.

In traditional models, exercise is framed as cardiovascular conditioning, strength building, and endurance progression. But for patients with autonomic impairment, exercise is a hemodynamic stress test, not a fitness routine.

When preload is inadequate, stroke volume impaired, baroreflex timing inconsistent, and sympathetic bandwidth limited, exercise does not strengthen the body—it destabilizes the system.

Rehab must therefore follow physiologic logic, not generic fitness advice.

This chapter outlines a clinician-centered model for mechanism-driven rehabilitation, preventing post-exertional deterioration, maximizing functional gain, and avoiding the predictable "exercise → crash → setback" cycle.

Section 1—The Fundamental Problem: Upright Exercise Requires Hemodynamic Integrity

Traditional exercise assumes that venous return is adequate, stroke volume increases appropriately, baroreflex sensitivity adjusts to maintain perfusion, sympathetic tone augments output, and vagal withdrawal is proportional and adaptive. In autonomic dysfunction, these assumptions collapse.

Common physiologic barriers include reduced preload, impaired venoconstriction, splanchnic pooling, small fiber-mediated vasodilation failures, hyperadrenergic overcompensation, delayed baroreflex response, and impaired chronotropic control.

Expecting patients to "walk more" or "go to the gym" under these conditions is physiologically incoherent.

Section 2—The Four-Phase Model of Autonomic Rehabilitation

Rehabilitation must begin where the physiology allows, not where generic exercise guidelines begin.

Phase 1—Horizontal Conditioning
This phase reduces gravitational load and allows:
- Improved venous return
- Reduced HR variability
- Stroke volume stabilization
- Enhanced tolerance of metabolic demand

Interventions:
- Recumbent cycling
- Rowing machines
- Swimming (when safe and appropriate)

- Supine resistance work
- Controlled breath training

Clinicians should counsel patients that horizontal conditioning is not "easy exercise"—it is foundational cardiovascular stabilization.

Phase 2—Semi-Recumbent Progression
As horizontal tolerance improves, patients advance to:
- Semi-recumbent cycling
- Angled rowing
- Low-load seated resistance training

Physiologic goals:
- Progressive loading of baroreflex latency
- Controlled sympathetic adaptation
- Gradual recruitment of lower extremity muscle pump without orthostatic collapse

This phase bridges gravitational physiology and functional gain.

Phase 3—Upright Reintroduction
Upright exercise is introduced only after hemodynamic stability is evident in lower-load phases.

Interventions:
- Treadmill walking
- Elliptical training
- Upright cycling
- Resistance training in standing positions

Clinician guidance:
- Start with short intervals
- Avoid prolonged static standing
- Monitor HR, RPE*, and symptom trajectory
- Maintain cooling strategies where appropriate

*RPE (rate of perceived exertion) helps contextualize heart rate when autonomic responses are unreliable.

Phase 4—Functional Integration
Patients transition from structured exercise to life-integrated activity:
- Walking at functional speed
- Stair tolerance
- Occupational mobility
- Low-intensity recreational activity

This is not equivalent to athletic training. It reflects the physiologic ceiling of accessible autonomic performance.

Section 3—Understanding the Crash: Why Overexertion Happens

Clinicians should recognize the pattern of:
1. Modest improvement
2. Overconfidence
3. Increased exertion
4. Physiologic crash
5. Reduced function for days

This pattern is not behavioral—it is biological.

Mechanisms Behind Post-Exertional Deterioration:
- Reduced cerebral perfusion during exertion
- delayed baroreflex reset
- Metabolic lag due to mitochondrial inefficiency
- Microvascular dysregulation
- Inflammatory signaling
- Autonomic afterload due to catecholamine excess

Rehab programs that ignore this physiology are designed to fail.

Section 4—Monitoring What Matters: Metrics Clinicians Should Use

Clinicians must track objective physiologic markers rather than relying on subjective descriptions of fatigue.

Recommended Measures:
- HR response patterns
- HR recovery
- Variability across intervals
- BP trends (if available)
- Temperature sensitivity
- Exertional symptom onset timing
- Delayed deterioration (12–72 hr surveillance)

This data helps clinicians correctly dose rehabilitation.

Section 5—Preload Conditioning as a Foundational Component

Functional improvement is impossible when circulating volume is insufficient.

Clinician-Directed Preload Optimization:
- Counsel adequate fluid and sodium intake
- Recommend compression garments
- Prescribe mineralocorticoids or vasoconstrictors when indicated
- Assess for splanchnic pooling
- Coordinate hydration protocols around exercise sessions

Exercise cannot overcome inadequate preload. Preload enables exercise.

Section 6—Temperature and Environment as Rehab Variables

Environmental conditions meaningfully change exercise tolerance.

Best Practices for Clinicians:
- Encourage climate-controlled settings
- Recommend cooling tools for heat-sensitive phenotypes
- Modify exercise dose based on thermal burden
- Anticipate lower tolerance in humidity or warm environments

Environmental mismanagement leads to physiologic instability, not psychological discomfort.

Section 7—Rehab Tailored to Phenotype

Not all autonomic dysfunction is created equal. Rehab must match phenotype.

Preload Failure:
- Rely heavily on horizontal conditioning
- Slower progression to upright
- Strong emphasis on compression and hydration

Hyperadrenergic Physiology:
- Avoid sympathetic over-activation
- Emphasize paced breathing
- Gradual intensity increases
- Monitor BP variability

Neuropathic Presentations:
- Incorporate temperature control
- Encourage controlled resistance training
- Anticipate delayed HR responses

Mixed Phenotype:
- Require highly individualized dosing
- Prioritize minimizing multi-system load

Section 8—Returning to Activity: Functional Goals

Functional reintegration is not about athletic milestones—it is about walking without orthostatic deterioration, tolerating daily tasks, regaining predictable physiologic responses, and reducing the frequency and severity of crashes. Physiologic capacity determines function, not motivation.

Science Snapshot—Mechanistic Layers of Autonomic Rehab

Mechanism	Rehab Target	Clinical Goal
Preload failure	Horizontal conditioning	Stabilize stroke volume
Baroreflex lag	Semi-recumbent work	Improve timing and adaptability
Sympathetic excess	Controlled pacing	Reduce NE over-recruitment

Mechanism	Rehab Target	Clinical Goal
Small fiber dysfunction	Temperature regulation	Minimize vasodilatory burden
Mitochondrial lag	Gradual dosing	Prevent post-exertional crashes

Translation—What This Chapter Really Means

Autonomic rehab is not "regular exercise at a slower pace." It is a mechanistic retraining of cardiovascular and neurologic responses that must follow predictable physiologic rules. Starting upright before the body is ready is not a sign of motivation; it's a trigger for relapse. Progress happens when rehab respects the biology of the condition.

Tilt Tip—The Clinician's Best Strategy

Clinicians dramatically improve outcomes when they validate the physiologic difficulty of exercise, tailor rehab to phenotype, monitor response over days—not just minutes, adjust dose based on symptoms and function, and emphasize progression through physiologic stability rather than speed. A stable foundation supports real improvement; overexertion sabotages it.

CHAPTER 21
THE AUTONOMIC TEST REPORT AS A WEAPON: HOW TO INTERPRET, INTEGRATE, AND ACTUALLY USE THE DATA

Because an autonomic report is only as powerful as the clinician reading it—and far too many reports languish unread in electronic medical record purgatory.

Introduction—The Problem Isn't the Testing: It's the Interpretation

Autonomic testing provides extraordinarily detailed physiologic information.

But in many clinical settings, the report is skimmed, misunderstood, ignored, misfiled, or dismissed as "normal" because the numbers weren't dramatic enough.

Meanwhile, patients remain symptomatic while their report quietly documents the physiology behind their dysfunction.

This chapter reframes the autonomic test report as a clinical weapon—a mechanistic tool that guides diagnosis, therapy, rehabilitation, and prognostication when interpreted with precision.

Section 1—The Anatomy of an Autonomic Test Report

A high-quality autonomic report should present:

1. Baseline Measures: HR, BP, respiratory pattern, temperature trends, sweat function baselines, and HRV metrics. These form the physiologic foundation upon which all subsequent responses are layered.
2. Challenge Responses: deep breathing, Valsalva maneuver, tilt/stand testing, and sudomotor evaluation. These expose latent deficits that resting measures conceal.
3. Pattern Recognition: The clinician's role is not simply to read values. It is to detect preload failure signatures, exaggerated sympathetic responses, blunted vagal control, baroreflex timing abnormalities, splanchnic redistribution evidence, and peripheral neuropathic patterns.
4. Integrative Summary: A report that ends with "consistent with dysautonomia" is incomplete. A report that ends with mechanistic classification is clinically transformative.

Section 2—Baseline Data: The Clues Hidden in Plain Sight

Baseline measures often hold the first signs of autonomic instability.

Key Indicators Clinicians Should Recognize:
- Mildly elevated HR may indicate low stroke volume
- Low HRV suggests poor vagal modulation
- Borderline low BP signals limited autonomic reserve
- Temperature asymmetry suggests small fiber dysfunction
- Resting tachycardia may signal chronic preload insufficiency

Baseline abnormalities are often dismissed because they are subtle. Subtlety is physiology's favorite disguise.

Section 3—Challenge Responses: When the Truth Reveals Itself

Autonomic challenges unmask what the baseline tries to hide.

A. Deep Breathing Test

Interpreting cardiovagal control requires attention to the amplitude of HR oscillation, cycle-to-cycle variability, and timing synchrony.

Reduced amplitude = impaired vagal modulation. Excessive variability = unstable autonomic gain.

B. Valsalva Maneuver

This test provides a window into sympathetic integrity, baroreflex timing, vascular compliance, and vagal rebound.

Common patterns clinicians miss:
- Depressed phase II BP compensation = impaired sympathetic vasoconstriction
- Exaggerated phase IV overshoot = hyperadrenergic physiology
- Blunted HR drop post-release = vagal insufficiency

The Valsalva is essentially a cardiovascular lie detector.

C. Tilt/Stand Testing

Tilt/Stand remains the most underutilized and misunderstood autonomic evaluation.

Mechanics clinicians should interpret:
- HR rise relative to BP stability
- Pulse pressure narrowing
- Cerebral perfusion markers (if available)
- Baroreflex oscillations
- Timing of instability

A tilt/stand curve is a narrative. Clinicians must read it from beginning to collapse—not just at the endpoints.

D. Sudomotor Testing

Small fiber assessment is often the missing link in explaining heat intolerance, inconsistent vasomotor responses, and reduced preload due to impaired cutaneous compensation. Sweat dysfunction is not cosmetic—it is physiologic.

Section 4—Recognizing the Dominant Failure Mode

Clinicians must determine which mechanism drives the patient's instability.

The Five Core Failure Modes:
1. Preload Failure
2. Hyperadrenergic Activation
3. Neuropathic Autonomic Impairment
4. Baroreflex Timing Dysfunction
5. Mixed Physiology

Reports should be interpreted through this lens, not through simplistic labels.

1. Preload Failure Signature
Key findings:
- Excessive HR rise
- Preserved or mildly reduced BP
- Narrow pulse pressure
- Low end-tidal CO_2 (when applicable)
- Delayed recovery after tilt

This phenotype requires volume-first strategies.

2. Hyperadrenergic Physiology
Look for:
- Exaggerated HR rise
- Elevated BP or late-tilt hypertensive response
- Tremor or sympathetic activation signs
- Valsalva phase IV overshoot

This phenotype demands careful modulation of sympathetic output.

3. Neuropathic Autonomic Patterns
Key indicators:
- Blunted sympathetic vasoconstriction
- Poor phase II Valsalva response
- Abnormal QSART or sudomotor deficits
- Reduced HRV
- Temperature dysregulation

This phenotype reflects impaired efferent autonomic pathways.

4. Baroreflex Dysfunction

Indicators:

- Inappropriate HR response relative to BP
- Exaggerated BP swings
- Delayed response timing
- Unstable oscillations during tilt

This requires timing and rehabilitation-focused strategies.

5. Mixed Physiology

This is the most common presentation.

Reports may show:

- Preload failure features
- Plus neuropathic deficits
- Plus hyperadrenergic compensation

Clinicians must resist the temptation to assign a single simplified label.

Section 5—Writing a Clinically Meaningful Report Summary

A clinically useful summary connects:

- Baseline data
- Challenge responses
- Pattern recognition
- Mechanistic classification
- Treatment implications

Example of what NOT to write:

- "Findings consistent with dysautonomia."

Example of what clinicians SHOULD write:
- "Findings indicate preload failure with secondary hyperadrenergic compensation, reduced vagal modulation, and impaired sudomotor response, consistent with mixed autonomic dysfunction."

That is a summary that leads to actual treatment.

Section 6—Turning the Report into a Clinical Plan

The test report is a diagnostic tool—not a decorative PDF.

Clinicians should use results to stratify phenotypes, tailor pharmacologic therapy, guide rehabilitation starting points, determine environmental sensitivity, predict functional limitations, justify accommodations, and coordinate multidisciplinary care.

Interpretation drives intervention.

Section 7—Communicating Findings Across Specialties

Autonomic specialists often need to coordinate with cardiology, neurology, gastroenterology, rheumatology, psychiatry, and endocrinology.

A mechanistic summary helps other clinicians contextualize symptoms without defaulting to psychogenic attributions. A concise report that names the mechanism prevents years of misinterpretation.

Science Snapshot—Mechanisms Revealed by Testing

Finding	Mechanistic Interpretation	Clinical Implication
Rapid HR rise, stable BP	Preload failure	Volume-first therapy
High BP during tilt	Hyperadrenergic activation	Sympathetic modulation
Blunted Valsalva phase II	Neuropathic dysfunction	Assess small fibers
Narrow pulse pressure	Low stroke volume	Compression, volume
Exaggerated phase IV	Hyperadrenergic spike	Controlled NE modulation
Absent sweat response	Small fiber involvement	Temperature management

Translation—What This Chapter Really Means

An autonomic test isn't a yes/no exam—it's a map of how your body manages blood flow, heart rate, pressure, and temperature. Clinicians use this map to identify which system isn't responding correctly and tailor treatment. When interpreted correctly, the report becomes a blueprint, not a mystery.

Tilt Tip—How Clinicians Use This Information Effectively

The strongest clinical outcomes occur when clinicians identify the dominant failure mode, explain findings in plain language, connect physiology to daily function, use the report to justify treatment and rehab strategies, and revisit results as patients progress.

Mechanism drives clarity. Clarity drives treatment. Treatment drives improvement.

CHAPTER 22
ACTION MAPS & CLINICAL RAPID REFERENCE: TURNING MECHANISMS INTO MEDICAL DECISIONS

Because clinicians shouldn't have to solve a physiologic murder mystery every time a dysautonomia patient walks in.

Introduction—Why Action Maps Matter

Autonomic dysfunction is not a single entity but a pattern-based physiologic landscape. Most patients present with overlapping mechanisms, inconsistent compensatory responses, and context-dependent deterioration.

Clinicians need a system that simplifies complex physiology, aligns symptoms with mechanisms, guides rapid treatment decisions, avoids misattribution, supports phenotype-based management, and prevents the "generic advice trap"

Action Maps do exactly that. They transform complicated autonomic data into clinically actionable pathways—a structured framework for immediate interpretation, intervention, and patient counseling.

This chapter provides a clinician-centered toolkit for rapid assessment and management decisions based on physiologic patterns.

Section 1—The Problem with Symptom-Centered Approaches

Autonomic dysfunction commonly presents with nonspecific multisystem symptoms: dizziness, tachycardia, fatigue, gastrointestinal complaints, heat intolerance, cognitive dysfunction ("brain fog"), weakness, and palpitations.

Symptoms do not reveal mechanisms. Mechanisms reveal mechanisms.

Action Maps shift clinical thinking from "What does the patient feel?" to "What underlying physiologic failure explains the presentation?" This prevents diagnostic drift and inappropriate management.

Section 2—The Action Map Framework

The Action Map system organizes autonomic dysfunction into four core domains:

1. Volume and Preload Status
2. Autonomic Modulation (sympathetic/vagal balance)
3. Neurovascular Control
4. External Load (environmental or behavioral)

Each domain contributes independently to instability. Together, they create the patient's lived physiologic reality. Clinicians can rapidly identify which domain is failing and match it to the appropriate intervention.

Section 3—Action Map 1: Volume & Preload Failure

This is the most common and most underdiagnosed contributor to autonomic instability.

Hallmark Clues:

- Standing intolerance
- Tachycardia out of proportion to effort
- Narrow pulse pressure
- Morning symptom clustering
- Post-meal or post-heat crashes
- Exercise intolerance despite motivation

Mechanistic Drivers:

- Inadequate circulating volume
- Venous pooling (lower extremity or splanchnic)
- Impaired vascular tone
- Small fiber dysfunction

Clinical Actions:

- Optimize fluids and sodium
- Prescribe compression strategically (calf → thigh → abdomen as needed)
- Consider mineralocorticoids or vasoconstrictors
- Integrate preload-focused rehab
- Assess for splanchnic dominance

When to Escalate:

- Persistent low stroke volume signals
- Recurrent morning collapses
- Syncope despite conservative measures

Section 4—Action Map 2: Sympathetic Overdrive & Vagal Under-Recruitment

Autonomic control depends on a delicate balance between sympathetic output and vagal modulation.

When sympathetic tone dominates or vagal tone collapses, patients exhibit:

- Excessive tachycardia
- Blood pressure volatility
- Sensory hypersensitivity
- Sleep fragmentation
- Exaggerated responses to stressors

Mechanistic Contributors:

- Chronic compensatory overactivation due to preload failure
- Alpha-adrenergic receptor hypersensitivity
- Baroreflex gain abnormalities
- Chronic inflammatory signaling
- Environmental overstimulation

Clinical Actions:

- Evaluate for underlying preload failure and correct first
- Consider low-dose beta-blockers or ivabradine when indicated
- Incorporate paced breathing protocols
- Teach controlled exposure to avoid sympathetic bandwidth depletion
- Support sleep structure and circadian stability

When to Escalate:
- BP spikes during tilt
- Persistent tremor or hyperadrenergic states
- Evidence of baroreflex maladaptation

Section 5—Action Map 3: Neurovascular Instability & Baroreflex Dysfunction

This map focuses on clinicians encountering:
- Exaggerated BP swings
- Orthostatic hypotension not explained by simple volume mechanics
- Delayed HR/BP coordination
- "Roller coaster" tilt-table patterns

Mechanistic Drivers:
- Impaired baroreceptor sensitivity
- Timing delays in afferent/efferent signaling
- Neuropathic lesions affecting vascular tone
- Endothelial dysfunction
- Medication interactions

Clinical Actions:
- Adjust medications affecting vascular response
- Recommend structured autonomic rehab for baroreflex training
- Manage triggers that increase baroreflex demand (heat, long standing, abrupt posture changes)
- Coordinate care with cardiology if BP volatility is severe

Red Flags:
- Syncope without prodrome
- BP swings >40 mmHg during tilt
- Exertional hypotension

Section 6—Action Map 4: External Physiologic Load (The "Context Map")

Environmental and situational factors dramatically modify autonomic stability.

Key Load Categories:
- Heat
- Humidity
- Meals
- Altitude
- Stress
- Sensory overload
- Prolonged standing
- Extended travel

These factors don't create autonomic dysfunction—they magnify existing deficits.

Clinical Actions:
- Identify patient-specific load profiles
- Provide anticipatory guidance (clinician-facing, not instruction scripts)
- Integrate environmental control into rehab phase progression
- Encourage workplace or school accommodations where physiologically justified

When to Anticipate Instability:
- After large meals
- During heat waves
- In airports or crowded venues
- After rapid position changes

Section 7—Cross-Map Integration: The Real-World Clinical Skill

Most patients exhibit two or more Action Maps simultaneously.

Clinicians must determine:
- The dominant mechanism
- The secondary modifiers
- The contexts that amplify instability
- The interventions that have the most leverage

Example Patterns:
- Preload failure + sympathetic excess
- Neuropathic dysfunction + baroreflex timing delays
- Splanchnic pooling + environmental heat load
- Mixed phenotype + sensory overload

A single "POTS" label does not capture this complexity. Mechanistic mapping does.

Section 8—Rapid Reference Tables for Clinical Use
Preload Failure: Quick Guide

Clue	Mechanism	Action
Tachycardia with narrow PP	Low SV	Preload optimization

Clue	Mechanism	Action
Morning worsening	Overnight hypovolemia	Salt + AM fluids
Post-meal crash	Splanchnic pooling	Smaller meals
Tilt tachycardia without hypotension	Central hypovolemia	Compression

Sympathetic Overdrive: Quick Guide

Clue	Mechanism	Action
Tremor, palpitations	Excess NE	Consider sympathetic modulation
BP spike on tilt	Hyperadrenergic state	Low-dose beta-blocker
Sensory overload	LC*-mediated activation	Environmental modulation

*LC = locus coeruleus.

Baroreflex Dysfunction: Quick Guide

Clue	Mechanism	Action
BP swings	Timing lag	Rehab with graded exposure
Syncope without warning	Baroreflex underperformance	Consider further cardio workup
HR/BP mismatch	Impaired gain	Adjust pharmacology

External Load: Quick Guide

Context	Mechanism	Action
Heat	Vasodilation	Cooling strategies
Meals	Redistribution	Structured intake

Context	Mechanism	Action
Travel	Preload disruption	Compression, hydration
Humidity	Thermal burden	Climate control

Section 9—Using Action Maps to Justify Accommodations or Interventions

Mechanism-based explanations help clinicians articulate medical necessity for work/school accommodations, travel modifications, temperature-controlled spaces, pacing modifications, rehab adjustments, and medication changes.

Action Maps provide clarity that insurers, employers, and other physicians recognize as evidence-based.

Section 10—Case Pattern Examples

Example 1—Preload Failure Dominant
Baseline: tachycardia, narrow PP
Tilt: exaggerated HR rise
Sudo: reduced secretion
Load: heat sensitive
Dominant Map: Volume
Treatment Focus: volume + compression + horizontal rehab

Example 2—Hyperadrenergic Dominant
Tilt: hypertensive response
Valsalva: exaggerated phase IV
Symptoms: tremor
Dominant Map: sympathetic excess
Treatment Focus: tone modulation + baseline preload optimization

Example 3—Mixed Mechanism
Findings: preload failure + neuropathic features
Symptoms: heat and meal triggers
Dominant Map: two-map interaction
Treatment Focus: dual-strategy plan

Science Snapshot—Action Maps as Clinical Logic

Action Map	Core Mechanism	Treatment Leverage
Preload	Low volume	Volume + compression
Sympathetic	NE excess	Tone modulation
Neurovascular	Baroreflex lag	Rehab + medication
External Load	Environmental stress	Anticipatory strategy

Mechanism dictates intervention. Action maps reveal mechanisms.

Translation—What This Chapter Really Means

Action Maps help clinicians understand which parts of the autonomic system are struggling. Instead of treating symptoms, clinicians target the underlying physiology—whether that's low blood volume, high adrenaline, nerve signaling problems, or environmental factors that push the system too hard.

Tilt Tip—The Clinician's Edge

Clinicians consistently achieve better outcomes when they identify the dominant mechanism using Action Maps, treat based on physiology (not labels), anticipate environmental load, and adjust over time as the patient's reserve increases.

This chapter isn't a list of tips—it's a diagnostic and therapeutic framework.

CHAPTER 23
REAL-WORLD ADAPTATION & SAFETY PROTOCOLS: CLINICIAN-GUIDED STRATEGIES FOR PREVENTING PHYSIOLOGIC AMBUSH

Because autonomic dysfunction doesn't care about clinical office hours—it attacks in grocery stores, hallways, showers, airports, dining rooms, and any other location patients can't reasonably avoid.

Introduction—Why Real-World Function Deserves Clinical Attention

Most clinicians underestimate how profoundly daily activities challenge impaired autonomic physiology. The clinic visit lasts 20–30 minutes (maybe); the patient's physiologic battle lasts every waking hour.

This chapter reframes "survival strategies" not as coping mechanisms but as clinician-guided adaptations grounded in autonomic physiology, hemodynamic load, environmental interaction, safety consideration, and functional recovery principles.

The objective is not to teach patients how to "push through" daily tasks—it is to teach clinicians how to reduce avoidable physiologic stress so that underlying mechanisms can stabilize through treatment and rehabilitation.

Autonomic dysfunction will always exploit real-world gaps. Clinicians can close those gaps.

Section 1—Why Real-World Function Is a Clinical Issue

Clinicians are trained to focus on laboratory results, imaging, and controlled clinical tests. However, autonomic instability often emerges in unpredictable environments—hot showers, crowded stores, standing at counters, post-meal periods, long conversations while standing, busy airports, and noisy workplaces.

These scenarios expose the autonomic system to stacked physiologic demands that cannot be captured in a clinical exam. Clinicians must guide patients through these realities—not because patients are fragile, but because the environments are physiologically adversarial.

Section 2—Standing Tasks: The Static Load Problem

Standing is not a neutral posture for individuals with impaired autonomic regulation. Static upright positioning creates a cascade of hemodynamic disadvantages.

Mechanisms:
- Up to 1 liter of blood pools in the legs and abdomen
- Lower extremity muscle pump is inactive
- Stroke volume declines
- HR rises to compensate
- Baroreflex demands increase
- Cerebral perfusion becomes vulnerable

Clinicians should recognize that "standing to perform tasks" is a medical stressor, not a lifestyle inconvenience.

Clinical Guidance:
- Recommend seated alternatives for prolonged tasks
- Encourage structured breaks from static standing

- Use compression to limit venous pooling
- Coordinate with occupational therapy for task modification

This improves safety and preserves functional capacity for more meaningful activities.

Section 3—Showering & Bathing: The Heat + Standing + Steam Triple Hit

Showering challenges autonomic stability through three simultaneous mechanisms:

1. Heat-induced vasodilation
2. Steam-induced impaired thermoregulation
3. Standing-induced venous pooling

These factors create a predictable environment for:
- Tachycardia
- Dizziness
- Hypotension
- Near-syncope

Clinical Guidance:
- Recommend cooler water and seated shower setups
- Encourage pre- and post-shower hydration when appropriate
- Discuss environmental ventilation (fans, ventilation fans, cooler rooms)
- Reinforce that shower intolerance is physiologic, not psychological

Shower modifications fall under medical safety optimization, not "comfort adjustments."

Section 4—Grocery Stores: The Perfect Storm of Physiologic Stress

Grocery stores combine fluorescent lighting, unpredictable temperatures, prolonged standing, crowding, decision-making demand, sensory overload, and heat zones near bakery or deli sections.

This makes them an unappreciated high-risk environment for autonomic instability.

Mechanisms:

- Sympathetic bandwidth overload
- Impaired vasomotor compensation
- Intermittent heat exposure
- Prolonged upright activity

Clinical Guidance:

- Counsel pacing in high-sensory environments
- Endorse environmental modifications such as supportive devices (carts, mobility aids)
- Normalize limitation as physiologic load management

Section 5—Post-Meal Vulnerability: Splanchnic Redistribution in Action

Meals redirect blood flow to the digestive tract, often triggering tachycardia, reduced cerebral perfusion, fatigue, thermal sensitivity, and prolonged recovery periods.

Mechanisms:

- 25–35% increased splanchnic blood flow
- Insulin-mediated vasodilation

- Reduced venous return
- Worsening of preload failure

Clinical Guidance:
- Recommend rest periods post-meal
- Encourage structured hydration around meals
- Highlight meal composition's physiologic impact

These are not dietary preferences—they are medically relevant hemodynamic considerations.

Section 6—Conversations While Standing: The Cognitive Load Trap

Patients often destabilize during prolonged standing conversations due to increased cognitive demand, reduced muscle pump activity, compounding baroreflex load, and splanchnic pooling if after meals.

This combination can precipitate significant instability.

Clinical Guidance:
- Suggest clinicians document the need for seating during conversations or meetings
- Reinforce that this is a cognitive-autonomic interaction, not a social issue

Section 7—Travel: A Multisystem Autonomic Stress Test

Airports, train stations, and long car rides reflect the highest density of environmental triggers in daily life.

Mechanisms:
- Prolonged upright posture
- Heat zones

- Dehydration
- Sensory overload
- Altitude changes
- Restricted movement

Clinical Guidance:
- Advise compression during travel
- Counsel structured hydration
- Endorse pacing and strategic breaks
- Provide documentation for accommodations where physiologically warranted

Travel is a clinical scenario, not just a lifestyle choice.

Section 8—Environmental Heat & Humidity: Predictable Physiologic Collapse

Heat amplifies vasodilation, blood pooling, sympathetic strain, fluid loss, and BP instability. Humidity further impairs evaporative cooling, overwhelming thermoregulatory mechanisms.

Clinical Guidance:
- Support access to climate-controlled environments
- Provide medical validation for cooling devices
- Document heat-sensitive physiology as part of disability or work accommodation assessments

Section 9—The Role of Clinician-Guided Adaptation

Autonomic instability is highly sensitive to environment and activity.

Clinicians improve outcomes when they teach environmental risk recognition, support adaptive strategies, incorporate real-world physiology into treatment planning, normalize functional modification as a therapeutic tool, and emphasize mechanistic explanations rather than behavioral framing.

Patients do not need "motivation"—they need appropriate physiologic dosing. This is a medical principle, not a mindset shift.

Section 10—Safety Considerations Clinicians Must Address

1. Fall Risk: driven by hypotension, sudden tachycardia, and delayed baroreflex response.
2. Syncope Risk: particularly during heat exposure, bathing, static standing, and prolonged travel.
3. Overexertion Risk: most relevant during meal preparation, chores, housework, and community activities.

Safety counseling is a clinical responsibility.

Science Snapshot—Real-World Load Components

Real-World Trigger	Mechanism	Physiologic Outcome
Shower	Heat + steam + standing	Tachycardia, hypotension
Grocery store	Sensory load + standing	Sympathetic depletion
Meals	Redistribution	Reduced preload
Travel	Upright time + stress	Hemodynamic instability
Conversations	Cognitive load + standing	Cerebral hypoperfusion

Translation—What This Chapter Really Means

Daily activities can create big physiologic demands when the autonomic system is impaired. The strategies clinicians recommend aren't about limitations—they're about matching your environment and activity level to what your physiology can safely manage. Real-world adaptation is part of treatment, not an admission of defeat.

Tilt Tip—How Clinicians Maximize Patient Safety

The best clinical outcomes happen when clinicians anticipate real-world triggers, personalize environmental guidance, integrate adaptation into treatment plans, validate that daily activities can be as physiologically demanding as exercise, and clearly document functional limitations when needed.

Helping patients navigate the "battlefield of daily life" is not optional—it is therapeutic.

CHAPTER 24
FUTURE DIRECTIONS OF AUTONOMIC MEDICINE

Because the autonomic nervous system is finally tired of being ignored, mischaracterized, and treated like the medical world's weird cousin.

Introduction—Welcome to the Renaissance (Finally)

For decades, autonomic medicine has lived in a strange limbo: too complex for casual clinicians, too niche for mainstream cardiology, too physiologic for neurology, too mysterious for research funding, and too poorly understood to attract large clinical trials.

And yet dysautonomia affects millions of patients, multiple organ systems, every domain of daily life, and every specialty that pretends it doesn't exist.

But the field is shifting—fast.

The next decade will bring new diagnostics, new therapeutics, advanced rehab models, precision phenotyping, mechanistic classifications, integrated care systems, and actual clinical guidelines that don't read like someone fell down an internet rabbit hole at 2 a.m.

This chapter shows where autonomic medicine is going—and why it's about to get much, much better.

Section 1—The Four Pillars of the Autonomic Future

The future of autonomic medicine will be built on four foundational pillars:

1. Precision Mechanistic Diagnosis
2. Real-Time Physiologic Monitoring
3. Advanced Rehabilitation Science
4. Mechanism-Driven Therapies

Everything else builds on these pillars. Let's break them down.

Pillar 1—Precision Mechanistic Diagnosis
Autonomic medicine has traditionally relied on inconsistent testing, partial interpretations, superficial summaries, and "mystery physiology" excuses. The future moves toward quantified, mechanistic diagnostic clarity.

Emerging technologies include high-frequency HRV analytics with full spectral mapping, continuous hemodynamic monitoring, noninvasive small fiber assessment, and AI-based autonomic pattern recognition.

This shifts diagnosis from "please explain these symptoms" to "here is your physiologic signature."

Pillar 2—Real-Time Physiologic Monitoring
The autonomic nervous system changes from second-to-second. Old diagnostics captured a snapshot; the future captures the full physiologic documentary.

Future wearables will track pulse pressure, baroreflex timing, HRV frequency bands, temperature gradients, sweat chloride conductance, respiratory variability, and environmental-load integration.

This creates a real-time physiologic dashboard.

Pillar 3—Advanced Rehabilitation Science
Rehab will shift from "exercise when you can" to "graded autonomic rehabilitation with mechanistic progression."

Expect autonomic-specific PT training, recumbent-to-upright progression centers, vagal and baroreflex conditioning protocols, and mitochondrial rehabilitation frameworks.

Rehab becomes a formalized science—not a guessing game.

Pillar 4—Mechanism-Driven Therapies
Future interventions will finally reflect actual physiology rather than vague intuition.

Expect advancements in preload modulation, sympathetic stabilization, vagal enhancement, nerve restoration, microvascular and endothelial-targeted therapies, and baroreflex retraining devices.

Treatment becomes targeted, predictable, and based on mechanism.

Spotlight Section—Long COVID: The Catalyst That Dragged Autonomic Medicine Into the Future

Long COVID didn't politely *contribute* to autonomic medicine—it drop-kicked the field through the nearest wall and forced everyone to pay attention. A single virus did what 30 years of conferences, papers, and polite specialist warnings could not: it made dysautonomia unignorable.

Here's how:

1. Long COVID proved autonomic dysfunction was never rare—just under-diagnosed

The pandemic exposed what autonomic clinicians already suspected: an enormous portion of the population was quietly walking around with borderline preload failure, borderline sympathetic instability, borderline small fiber dysfunction, and borderline baroreflex inconsistency.

All it took was a major viral insult to shove borderline physiology straight into the danger zone. The dysfunction wasn't new—the volume just got turned up.

2. Long COVID triggered the autonomic research boom we'd been waiting decades for

Suddenly, large research teams started investigating viral effects on autonomic ganglia and the brainstem, endothelial and microvascular injury, immune-mediated small fiber damage, mitochondrial-autonomic interactions, persistent baroreflex dysregulation, and cardiovagal impairment.

Topics once buried in niche autonomic meetings became NIH-funded priorities. That's unprecedented.

3. Long COVID created the first large-scale population of mixed-mechanism dysautonomia

Patients stopped presenting in tidy boxes and began showing delightful combinations of preload failure, hyperadrenergic activation, neuropathic signatures, vagal impairment, and baroreflex latency.

Long COVID became the proof-of-concept that real physiology doesn't respect diagnostic categories.

4. Long COVID popularized real-time physiologic monitoring

Wearables went from "cute wellness gadget" to objective proof of physiologic symptoms, continuous HRV surveillance, BP and pulse-pressure estimation, environmental-load integration, and post-exertional instability mapping.

Real-time monitoring is standard now because long COVID demanded receipts.

5. Long COVID forced mainstream recognition of post-viral autonomic syndromes

Before COVID, post-viral dysautonomia was documented, known to specialists, and politely ignored.

The pandemic forced acknowledgment of small fiber injury, immune-driven sympathetic instability, endothelial dysfunction, mitochondrial lag, and persistent baroreflex distortion.

Post-viral syndromes are no longer exotic—they're everywhere.

6. Long COVID accelerated mechanistic classification

"POTS" officially outgrew its usefulness. The field now leans toward preload subtypes, neuropathic subtypes, inflammatory subtypes, hyperadrenergic subtypes, and mixed-phenotype labeling.

Diagnosis is shifting from symptom clusters to biologic categories.

7. Long COVID advanced autonomic rehabilitation science

Rehab teams learned quickly that exertional crashes require physiologic pacing, stroke volume recovery curves matter, horizontal-first rehab is mandatory, preload control directly affects exercise tolerance, and baroreflex lag must be trained—not ignored.
These insights apply far beyond long COVID—they elevate dysautonomia care universally.

8. Long COVID accelerated the push for autonomic centers of excellence

Demand exploded for tilt-table facilities, autonomic labs, interdisciplinary clinics, and remote monitoring programs.

That infrastructure continues to benefit dysautonomia patients long after COVID became context, not cause.

Science Snapshot—The Autonomic Future Matrix

Domain	Current State	Future State
Testing	Snapshot-based	Continuous mechanistic monitoring
Classification	Umbrella labels	Mechanistic phenotyping
Rehab	Generic exercise	Autonomic-specific conditioning
Treatment	Symptom-first	Mechanism-first
Data Use	Passive	AI-driven active modeling
Research	Limited	Multi-center coordinated
Clinical Access	Rare specialists	Widespread autonomic centers

Autonomic medicine is no longer drifting at the periphery of clinical care. It is becoming data-driven, mechanism-mapped, precision-engineered, and technologically augmented.

And thanks to long COVID's disruptive influence, the field is advancing faster than any other domain of physiology. The renaissance has begun.

Translation—What This Section Really Means

Long COVID didn't invent autonomic medicine—it just dragged it out of the basement and into the spotlight.

As millions suddenly developed preload instability, small fiber involvement, hyperadrenergic patterns, baroreflex dysfunction, and exertional vulnerability, the field finally gained funding, research momentum, diagnostic precision, and clinical legitimacy.

Autonomic medicine is entering a long-overdue renaissance. Long COVID was the accelerant.

Tilt Tip—How This Shapes the Future

You're witnessing the fastest evolution in autonomic physiology in history. The next decade will revolve around mechanisms illuminated by long COVID research. What we learn from post-viral dysautonomia will raise the standard of care for *all* autonomic patients. Understanding these patterns now places you well ahead of the coming wave of breakthroughs.

CONCLUSION
THE GRAVITY-DEFYING TRUTH

You made it to the end, which already puts you ahead of most clinicians who never got past the introductory definition of dysautonomia before deciding the autonomic nervous system was "too complicated" and pivoting to a career where every symptom magically becomes "psychogenic."

This book has taken you through the mechanics of autonomic dysfunction, the physiology behind every test, the failure modes that actually matter, the survival strategies for an uncooperative world, the treatment sequences that work, and the real future of autonomic medicine.

Now we tie it all together—not with vague motivational platitudes, but with the blunt truths the field desperately needs.

1. Dysautonomia is not mysterious. It's mechanistic.
Nothing in this book required magic, mysticism, or performance-art diagnostics. Your physiology is not unpredictable—it is consistent, logical, and reliably awful in the exact same ways every time gravity gets involved.

If you know the mechanism, you know the outcome. If you know the outcome, you know the treatment. If the treatment fails, you misidentified the mechanism.

That's it. That's the secret sauce.

The entire field would move forward ten years overnight if everyone stopped using acronyms as diagnoses and started identifying *why* the system fails.

2. Autonomic testing is underutilized, undervalued, and undersold. Most patients with dysautonomia never receive testing.

Those who do often get incomplete testing, poorly interpreted testing, contradictory summaries, and diagnostic labels that describe symptoms, not physiology.

You now know how to read these tests better than many clinicians.

You understand what Phase II means, why Phase IV matters, what HRV actually tells you, and why ESC is not a trivia point—it is the smoking gun of small fiber function.

You have gone from a passive recipient of medical impressions to an active interpreter of physiology. This is how the power imbalance flips.

3. Treatment fails when it's generic. Treatment works when it's targeted. If you learned nothing else, learn this—generic dysautonomia treatment helps almost no one.

Mechanism-directed therapy helps almost everyone:
- Preload failure needs volume.
- Vascular failure needs tone.
- Baroreflex failure needs stability.
- Small fiber dysfunction needs cooling, pacing, and nerve support.
- Hyperadrenergic physiology needs sympathetic modulation, not gallon-jug salt challenges.
- Treating POTS like one condition is like treating all chest pain with an antacid.

Precision is not optional.

4. The real world is a physiological gladiator arena.
Gravity is relentless. Heat is merciless. Meals are saboteurs. Travel is a multi-phase autonomic endurance test. Standing still is a crime against preload. And cognitive load while upright is the neurological equivalent of running 14 apps on a cell phone clinging to 3% battery.

But now you have survival protocols, pacing architecture, trigger-stack avoidance, environmental combat strategies, motion safety plans, and thermal warfare techniques.

You have a toolkit designed for the reality you actually live in—not the fantasy world in which clinicians imagine people sit quietly in air-conditioned rooms all day.

5. Autonomic medicine is changing—and you're ahead of it.
Most patients don't realize they're living five to ten years ahead of the healthcare system. You are one of them.

The field *will* evolve:
- Mechanistic diagnoses will replace umbrella acronyms.
- Wearables will replace guesswork.
- Neuromodulation will replace dismissive reassurance.
- Small fiber research will explode.
- Post-viral autonomic science will rewrite entire treatment models.
- Testing will become more accessible, more automated, and more precise.

You have already learned the language that future clinicians will be trained in.

6. Your autonomic system may be tilted, but you're not down.

Here's the real takeaway: dysautonomia isn't a personality flaw, a psychological quirk, a motivational deficit, or an inability to hydrate aggressively enough to impress a nephrologist.

It is physiology. It is real. It is measurable. It is actionable. It is survivable. And for many, it is improvable—with the right framework.

You now have that framework—the science, the tests, the mechanisms, the strategies, the action maps, and the future-facing roadmap.

You're no longer standing in the autonomic storm without a system. You *are* the system.

FINAL THOUGHT
THE LAST WORD FROM BOTH SIDES OF THE EXAM TABLE

(Because gravity never takes a day off.)

For Patients—You Didn't Imagine the Physiology. Medicine Just Didn't Measure It.

Let's clear something up: You were never "overreacting." You were under-measured.

Every time you stood up and your heart rate skyrocketed like it was late for a meeting, that was physiology. Every time a hot room made you dissolve into a puddle of neurological chaos, that was physiology. Every time a doctor shrugged at your symptoms like they were evaluating a weird noise in a rental car, that was physiology too.

You weren't "dramatic." You were hemodynamically accurate. And now? You're dangerous—an autonomic samurai warrior carved out of pure physiologic precision.

You know more about the autonomic nervous system than many people who took an oath to understand it. You can walk into an appointment and spot the difference between preload failure, vascular incompetence, baroreflex confusion, and a clinician who thinks "POTS is dehydration" with the same speed and precision.

You can explain your mechanism. You can analyze your triggers. You can predict your crashes. You can stabilize them faster than someone can say "Have you tried breathing exercises?" You didn't just survive dysautonomia—you reverse-engineered it.

Tilt happens. And now you tilt with intent.

For Clinicians—If You Think These Patients Are Complicated, Try Understanding the Physiology.

If you've made it to this point, let me be painfully clear—your patient isn't confusing. Your training was incomplete.

Nothing about this physiology is "mysterious." It only *feels* mysterious if your only autonomic education came from a cardiology lecture featuring one bullet point, a neurology rotation too busy diagnosing strokes, or a psych consult note that said "ruled out" without ruling out a single thing.

These patients aren't attention-seeking. They're physiology-seeking. They want mechanisms, not mood interpretations. They want targeted treatment, not stress-management pamphlets. They want someone who understands that vasodilation isn't a personality trait.

If you walk away from this book with one thing, let it be this:

When you measure the right signals and interpret them correctly, dysautonomia stops looking like chaos and starts looking like a system—a system with identifiable failure modes, predictable responses, and actionable pathways.

These patients don't need you to fix them with guesswork. They need you to meet them in the physiology. Tilt happens. Your job is to understand why. And now, finally, you can.

Khalid Saeed, D.O.
Baroreflex Whisperer | Autonomic Wrangler | Gravity Defiant

ABOUT THE AUTHOR
THE CLINICIAN WHO READ THE FINE PRINT

(Because apparently you can't bill the ANS for emotional damages.)

Dr. Khalid Saeed, D.O., is a board-certified physician who did what most doctors avoid at all costs: he paid attention to the autonomic nervous system. Not because the medical curriculum encouraged it (it didn't), or because residency made space for it (it definitely didn't), but because his own ANS eventually staged enough mutinies that ignoring it stopped being an option.

Like most clinicians, Dr. Saeed was originally taught that the autonomic nervous system "runs in the background"—a phrase that aged poorly once his own background processes began throwing error codes. So he did what any overeducated, sleep-deprived physician would do: he hunted down every overlooked mechanism, every buried physiology paper, and every test interpretation guideline that wasn't sabotaged by vague adjectives.

He quickly discovered a pattern: the nervous system wasn't the problem—the way medicine talks about it was. So he rewrote the conversation.

Armed with lived experience, clinical expertise, and a healthy disrespect for bad explanations, Dr. Saeed dismantled the idea that dysautonomia is confusing and rebuilt it into what it always should have been: a set of measurable mechanisms with predictable behavior and treatable failure modes. Patients appreciated this. The medical system tolerated it. Gravity filed a restraining order.

This book represents decades of clinical practice, patient advocacy, physiologic investigation, and personal "field testing" performed by a doctor who lives inside the very system he explains. It brings together the honesty of a patient, the precision of a clinician, and the snark of someone whose autonomic system has repeatedly reminded him that he is, in fact, not in charge.

When he's not teaching clinicians why blood pressure shouldn't behave like interpretive art, Dr. Saeed practices medicine, writes, organizes physiology into coherent frameworks, and lives with the strategic planning of someone whose ANS can be either helpful or hostile depending on the temperature of the room.

He currently lives, works, and biologically negotiates with gravity in Florida. Tilt happens. But it doesn't get the final say. Not anymore.

You can reach Dr. Saeed at TampaBayConciergeDoctor.com or on Instagram at dr.khalid.saeed

CREDITS

(Because even books about rebellious autonomic systems need a formal thank-you section.)

To the autonomic nervous system: Thank you for providing endless material by refusing to behave in any predictable, linear, textbook-approved manner. You are the reason this book exists, and also the reason the author owns more cooling devices than kitchen appliances.

To the patients who kept showing up even after being told, "All your tests look fine": You're the reason this field advances. You're also the reason many clinicians suddenly develop a search history suspiciously filled with phrases like "Why did my training skip an entire organ system?" at 2 a.m.

To the clinicians who didn't dismiss dysautonomia as "anxiety with flair": You are the heroes of this specialty. You will be rewarded in the next life with short clinic days, functioning EMRs, and cardiology colleagues who answer their pages on the first try.

To the clinicians who *did* dismiss dysautonomia as anxiety: Thank you for inadvertently motivating this book. Your contribution, though unintentional, was significant.

To the Valsalva maneuver: You complete disaster of a test. Thank you for revealing baroreflex dysfunction, sympathetic integrity, and exactly how annoyed a patient can look while bearing down.

To heat, humidity, and every environment that triggered symptoms during writing: Thank you for the live demonstrations. Truly immersive research.

To gravity: We get it. You're undefeated. There's no need to show off every time someone stands up.

To the people who insisted dysautonomia was "too complicated to explain," consider this book your polite correction.

To every chair, couch, and bed the author has strategically occupied while writing: You all carried this project. Literally.

To the readers, patients, clinicians, skeptics, curious humans, and the autonomically bewildered: Thank you for making it this far. You now know more about the autonomic nervous system than several committees that write national guidelines.

Tilt happens. Science helps. Snark sustains. And now you have all three.

REFERENCES

(Because at some point in history, a scientist measured all the things your doctor insists on eyeballing.)

Before we dive into the citations, let's establish one thing: autonomic medicine is not powered by opinions, inspirational slogans, or the eternal clinical classic "your tests are normal."

Everything in this book rests on decades of physiology, neurology, and cardiovascular research conducted by scientists who—unlike most clinical guidelines—actually measured things.

So here they are: the peer-reviewed breadcrumbs that prove your autonomic nervous system is not dramatic, confused, or "just anxious." It's physiologic. It's measurable. And it's been documented thoroughly enough that no clinician should still be guessing.

Below is a curated selection of the foundational evidence that anchors the mechanisms, interpretations, and testing principles throughout *Tilt Happens*.

I. CORE AUTONOMIC PHYSIOLOGY

Benarroch, E. E. (1993). The central autonomic network: Functional organization. *Neurology*, 43, 243–248.

Benarroch, E. E. (2012). *Autonomic Nervous System: Basic Anatomy and Physiology.* Continuum.

Benarroch, E. E. (2020). *The Central Autonomic Nervous System.* Mayo Clinic Proceedings.

Jänig, W. (2008). *Integrative Action of the Autonomic Nervous System.* Cambridge University Press.

Robertson, D., Biaggioni, I., Burnstock, G., Low, P. A., & Paton, J. F. R. (2012). *Primer on the Autonomic Nervous System* (3rd ed.). Academic Press.

Saper, C. B. (2002). Brainstem control of autonomic function. *Autonomic Neuroscience*, 98, 3–14.

Blessing, W. W. (1997). *The Lower Brainstem and Bodily Homeostasis*. Oxford University Press.

Low, P. A. (1997). *Clinical Autonomic Disorders* (1st ed.). Lippincott Williams & Wilkins.

Mathias, C. J., & Bannister, R. (2013). *Autonomic Failure* (5th ed.). Oxford University Press.

Hopkins, D. A. (1975). Autonomic pathways and central organization. *Journal of Comparative Neurology*, 161, 143–170.

II. BAROREFLEX, HR CONTROL, AND VALSALVA

Cerutti, C., Barres, C., & Paultre, C. (1994). Baroreflex modulation during Valsalva. *Clinical Autonomic Research*, 4(1), 5–12.

Fisher, J. P., & Young, C. N. (2021). Baroreflex role in cardiovascular health. *Hypertension*, 78, 1823–1832.

El-Sayed, H., & Hainsworth, R. (1995). Baroreceptor function during Valsalva. *Clinical Science*, 88, 15–20.

Levin, A. B. (1966). Valsalva maneuver and autonomic function. *American Journal of Cardiology*, 18, 90–99.

Borovikova, L. V., et al. (2000). Vagal anti-inflammatory pathway. *Nature*, 405, 458–462.

Mancia, G., & Mark, A. L. (1983). Arterial baroreflexes in humans. *Handbook of Physiology*, 755–793.

III. HEART RATE VARIABILITY (HRV)

Shaffer, F., & Ginsberg, J. P. (2017). HRV metrics and norms. *Frontiers in Public Health*, 5, 258.

Task Force. (1996). Standards of measurement and interpretation of HRV. *Circulation*, 93, 1043–1065.

Goldstein, D. S., Bentho, O., Park, M. Y., & Sharabi, Y. (2011). Low-frequency HRV is not sympathetic tone. *Clinical Autonomic Research*, 21, 133–143.

Sacha, J. (2014). Correcting HRV for heart rate. *Frontiers in Physiology*, 5, 168.

Boudreau, P., Yeh, W., & Dumont, G. (2012). HRV across sleep stages. *Sleep Medicine*, 13, 784–788.

Porta, A., et al. (2014). HRV methods in clinical research. *Philosophical Transactions of the Royal Society A*, 373, 2014–2023.

Lanfranchi, P., & Somers, V. (2002). Autonomic activity during sleep. *Journal of Applied Physiology*, 92, 2146–2156.

Thayer, J. F., Hansen, A. L., et al. (2009). Vagal tone and health. *Biological Psychology*, 74, 224–242.

IV. ORTHOSTATIC INTOLERANCE, POTS, AND TILT TESTING

Freeman, R., Wieling, W., Axelrod, F. B., et al. (2011). Consensus statement on POTS & OH. *Clinical Autonomic Research*, 21, 69–72.

Raj, S. R. (2013). Postural tachycardia syndrome. *Circulation*, 127, 2336–2342.

Bryarly, M., Phillips, L. T., Fu, Q., Vernino, S., & Levine, B. D. (2019). POTS: JACC review. *JACC*, 73, 1207–1228.

Fedorowski, A. (2019). POTS: Clinical overview. *Journal of Internal Medicine*, 285, 352–366.

Stewart, J. M. (2013). Sympathetic activation in POTS. *Autonomic Neuroscience*, 177, 69–73.

Lahr, B. D., et al. (2020). Interpretation of tilt-table testing. *Clinical Autonomic Research*, 30, 1–8.

Sheldon, R., et al. (2015). Syncope classification. *Circulation*, 132, 171–179.

Fu, Q., & Levine, B. (2018). Orthostatic intolerance in chronic disease. *Autonomic Neuroscience*, 215, 14–22.

Benditt, D. G., et al. (1996). Tilt-induced syncope. *Journal of the American College of Cardiology*, 28, 165–175.

V. NEUROCARDIOGENIC SYNCOPE AND BAROREFLEX FAILURE

Grubb, B. P. (2005). Neurocardiogenic syncope mechanisms. *JACC*, 46, 142–148.

Jacob, G., et al. (1997). Baroreflex failure. *NEJM*, 336, 1449–1455.

Biaggioni, I., & Robertson, D. (2002). Reflex syncope and autonomic failure. *Primer on the Autonomic Nervous System*, 425–431.

van Dijk, N., et al. (2006). Syncope epidemiology. *Neurology*, 66, 25–30.

VI. SUDOMOTOR FUNCTION AND SMALL FIBER NEUROPATHY

Sandroni, P., Opfer-Gehrking, T., & Low, P. A. (1999). Small fiber autonomic neuropathy. *Neurology*, 53, 1431–1438.

Illigens, B. M. W., & Gibbons, C. H. (2009). QSART: Principles and interpretation. *Autonomic Neuroscience*, 146, 35–46.

Gibbons, C. H., et al. (2010). Electrochemical skin conductance. *Muscle & Nerve*, 42, 456–462.

Devigili, G., et al. (2008). Diagnostic criteria for small fiber neuropathy. *Brain*, 131, 1912–1925.

Oaklander, A. L., et al. (2013). SFN in postinfectious states. *PAIN*, 154, 2284–2289.

Bakkers, M., et al. (2014). Small fiber neuropathy: A common underrecognized condition. *Annals of Neurology*, 75, 55–64.

Shy, M. E., et al. (2003). The autonomic nervous system in peripheral neuropathies. *Lancet Neurology*, 2, 356–365.

VII. HEAT PHYSIOLOGY, THERMOREGULATION, AND C-FIBERS

Crandall, C. G., & González-Alonso, J. (2010). Heat stress physiology. *Journal of Applied Physiology*, 109, 1–9.

Shibasaki, M., & Crandall, C. G. (2011). Sweat gland control mechanisms. *Autonomic Neuroscience*, 158, 82–90.

Nadel, E. R., et al. (1971). Skin temperature and thermoregulation. *Journal of Applied Physiology*, 31, 80–87.

Kenny, G. P., & Jay, O. (2013). Thermoregulatory physiology. *Comprehensive Physiology*, 3, 1689–1739.

Fellows, I. W., et al. (1988). C-fiber dysfunction & heat intolerance. *Neurology*, 38, 1677–1681.

VIII. POST-VIRAL DYSAUTONOMIA AND LONG COVID

Vernino, S., Stiles, L. E., & Drolet, H. (2021). Autonomic dysfunction in long COVID. *Nature Reviews Neurology*, 17, 1–12.

Dani, M., et al. (2021). Autonomic dysfunction in post-COVID states. *Clinical Medicine*, 21, e68–e72.

Goldstein, D. S. (2019). The extended autonomic system and viral dysautonomia. *Clinical Autonomic Research*, 29, 237–239.

Raj, S. R., et al. (2020). Viral triggers of autonomic disorders. *Autonomic Neuroscience*, 228, 102700.

Kemp, H. I., et al. (2020). Post-infectious small fiber neuropathy. *Pain Reports*, 5(4), e831.

IX. AUTOIMMUNE AUTONOMIC DISORDERS

Vernino, S., & Low, P. A. (2016). Autoimmune autonomic ganglionopathy. *Handbook of Clinical Neurology*, 133, 405–415.

Sene, D. (2018). Small fiber neuropathies: Autoimmune etiologies. *Current Opinion in Neurology, 31,* 569–579.

Gibbons, C. H., & Freeman, R. (2009). Antibodies in autonomic neuropathy. *Neurology, 72,* 2002–2008.

Klein, C. M. (2013). Autoimmune autonomic disorders. *Seminars in Neurology, 33,* 57–64.

Shibao, C., et al. (2013). Immune-mediated autonomic failure. *Autonomic Neuroscience, 177,* 62–68.

X. GI MOTILITY AND AUTONOMIC CONTROL

Camilleri, M. (2018). Autonomic control of GI motility. *Neurogastroenterology & Motility, 30,* e13210.

Karasawa, H., et al. (2013). Enteric neuropathy mechanisms. *Journal of Physiology, 591,* 4379–4389.

Sarna, S. K. (2010). Physiology of gastrointestinal motility. *Comprehensive Physiology, 3,* 919–962.

XI. EXERCISE PHYSIOLOGY AND AUTONOMIC DYSFUNCTION

Fu, Q., & Levine, B. D. (2018). Exercise intolerance in autonomic dysfunction. *Autonomic Neuroscience, 215,* 14–22.

Saltin, B., et al. (1998). Deconditioning vs. autonomic failure. *Circulation, 97,* 96–102.

Winker, R., et al. (2005). Exercise intolerance and autonomic patterns. *Medicine & Science in Sports & Exercise, 37,* 136–142.

Shibata, S., et al. (2012). Cardiovascular recovery and autonomic tone. *Journal of Physiology, 590,* 6475–6486.

XII. MEDICATION EFFECTS ON AUTONOMIC TESTING

Biaggioni, I., & Robertson, D. (2002). Drug effects on autonomic function. In *Primer on the Autonomic Nervous System,* 425–431.

Shibao, C., Okamoto, L., Stein, P. K., & Biaggioni, I. (2013). Medications in autonomic testing. *Autonomic Neuroscience, 176,* 92–99.

Jordan, J., et al. (2000). Central sympatholytics in autonomic disorders. *Journal of Clinical Hypertension, 2,* 326–332.

XIII. MECHANISM-DIRECTED THERAPY

Stewart, J. M. (2019). Mechanistic treatment of orthostatic intolerance. *Journal of Pediatrics, 210,* 12–22.

Garland, E. M., et al. (2015). POTS treatment beyond volume expansion. *Autonomic Neuroscience*, 215, 1–8.

Kizilbash, S. J., et al. (2014). Comprehensive management of autonomic disorders. *Pediatric Neurology*, 51, 1–13.

Grubb, B. P., & Karas, B. (1999). Management of chronic autonomic dysfunction. *Journal of Interventional Cardiac Electrophysiology*, 3, 185–196.

XIV. CEREBRAL PERFUSION AND COGNITIVE IMPACT

Ocon, A. J. (2013). CBF dysregulation in autonomic disorders. *Autonomic Neuroscience*, 177, 84–92.

van Campen, C., Rowe, P., & Visser, F. (2020). Reduced cerebral blood flow in ME/CFS and POTS. *Frontiers in Pediatrics*, 8, 600.

Stewart, J. M. (2012). Orthostatic cerebral perfusion. *Neurophysiology*, 29, 191–198.

XV. THERMAL, SENSORY, AND MICROVASCULAR INTEGRATION

Rowell, L. B. (1983). Human cardiovascular control in heat. *Handbook of Physiology*, 967–1024.

Minson, C. T., et al. (2001). Skin blood flow regulation. *Journal of Applied Physiology*, 91, 1619–1626.

Tjen-A-Looi, S. C., et al. (2005). C-fiber modulation and reflexes. *Journal of Physiology*, 564, 981–995.

XVI. GENERAL AUTONOMIC TESTING METHODS

Low, P. A. (2003). Laboratory evaluation of autonomic failure. *Autonomic Failure*, 287–307.

Gibbons, C. H. (2010). Autonomic testing basics. *Continuum*, 16, 4–19.

Malliani, A., Pagani, M., & Lombardi, F. (1991). Power spectral HRV interpretation. *Circulation*, 84, 482–492.

SUGGESTED READINGS & TOOLS

(Because knowing things is more effective than guessing, and your autonomic nervous system deserves better than vibes.)

This section exists for one reason: if you've made it through this book and still want more, you are officially the kind of person who reads autonomic literature voluntarily. Congratulations. You are now part of the 0.01% of humanity who understands that "your vitals were normal" is not a clinical interpretation, and you deserve resources that won't insult your intelligence.

Below is a curated list of books, papers, tools, apps, and diagnostic aids that actually help clinicians and patients navigate dysautonomia. No pseudoscience, no wellness glitter, no "just breathe deeper and see what happens." These are the sources that keep autonomic medicine anchored in reality.

I. Books Worth Reading (Because They Contain Actual Physiology)

1. *Clinical Autonomic Disorders* (Low & Benarroch)
The gold standard. If autonomic medicine were a religion, this would be the sacred text. Dense, technical, and absolutely worth it.

2. *Autonomic Failure* (Bannister & Mathias)
If you want to understand central autonomic pathways at a level that makes your colleagues uncomfortable, this book is your new best friend.

3. *Primer on the Autonomic Nervous System* (Robertson, Biaggioni, et al.)
Concise and readable. Should be required reading for anyone who uses the phrase "it's probably anxiety."

4. *Integrative Action of the Autonomic Nervous System* (Jänig)
A deep dive into how autonomic physiology coordinates survival. Not casual reading unless you consider neurophysiology casual.

5. *The Dysautonomia Project* (Molderings et al.)
More patient-accessible, but still grounded in science. A good bridge resource for clinicians learning the field.

6. *TILTED: A Medical Memoir of Dysautonomia and Other Horizontal Pursuits* (Saeed, K.)
An origin story, a rebellion, a science-backed saga of dysautonomia survival. Equal parts physiology, gallows humor, and useful rage. If *Tilt Happens* is the technical manual, *Tilted* is the narrative that explains exactly why the manual needed to exist.

II. Key Papers (Your Brain Will Thank You Later)
1. *HRV & Parasympathetic Physiology*
    - Task Force of the ESC & NASPE—HRV Standards
    - Shaffer & Ginsberg—HRV methodology and interpretation
    - Goldstein et al.—Why low-frequency HRV is not sympathetic tone

2. *Orthostatic Intolerance & POTS*
    - Raj—Circulation overview of POTS
    - Bryarly et al.—JACC POTS seminar
    - Freeman et al.—Consensus definitions for OH, POTS, and NMS

3. *Sudomotor Function & Small Fiber Neuropathy*
    - Gibbons & Illigens—Mechanisms behind QSART
    - Devigili et al.—Diagnostic criteria for small fiber neuropathy
    - Gibbons et al.—ESC and sweat gland innervation

Suggested Reading & Tools

4. Baroreflex & Valsalva Interpretation
- Cerutti et al.—Baroreflex modelling
- Fisher & Young—Baroreflex and cardiovascular stability

5. Long COVID & Post-Viral Dysautonomia
- Vernino et al.—Autonomic dysfunction after COVID
- Dani et al.—Clinical manifestations and testing pathways

These papers are not for light reading. They are for people who enjoy evidence, nuance, and heart rate variability graphs.

III. Tools Clinicians Should Actually Use
1. Autonomic Testing Suite
   - Includes HRV, deep breathing, Valsalva, sudomotor testing, QSART, orthostatic vitals, and beat-to-beat BP. If your clinic lacks this equipment, it is not an autonomic clinic.

2. Continuous Beat-to-Beat Blood Pressure Monitoring (Finapres or equivalent)
   - Stop relying on static cuff readings. Dysautonomia is a timing disorder; you need real-time data.

3. Electrochemical Skin Conductance Devices (Sudomotor Testing)
   - Pain-free, fast, and diagnostic. Should be in every neurology, cardiology, and integrative clinic.

4. HRV Analysis Tools
   - Kubios HRV
   - Elite HRV (consumer tier but still educational)
   - Oura, WHOOP, and Aura devices (helpful for trends, not diagnostics)

Use these tools for patterns, not vibes. HRV is a physiologic signal, not a personality score.

5. Ambulatory BP & HR Monitoring
- For catching real-world physiology instead of the polite, clinic-appropriate version.

IV. Apps & Digital Tools (Use Responsibly)
1. *Kubios HRV (Pro & Research Levels)*
Accurate HRV interpretation. Not the place for motivational quotes or pastel graphics.

2. *PoTS UK Symptom Tracker*
Useful for patient-tracked symptom patterns, especially flare-trigger identification.

3. *CARP Autonomic Testing PDFs & Algorithms*
Flowcharts clinicians will actually understand and—better yet—use.

4. *Journal-based Apps (Bear, Notion, Obsidian)*
For those who prefer structured physiologic logging instead of scribbling "felt terrible again" on the nearest napkin.

V. Patient-Facing Resources That Don't Spread Nonsense
1. *Dysautonomia International*
Comprehensive educational material, research updates, and advocacy.

2. *PoTS UK*
Clear, evidence-grounded explanations with guidelines clinicians should frankly plagiarize.

3. *The Dysautonomia Project*

A patient-friendly clinical overview that doesn't oversimplify mechanisms.

4. *ME/CFS & OI Clinical Guidelines (CDC and Johns Hopkins)*
Particularly valuable for understanding exertional intolerance and post-viral physiology.

VI. Tools for Building a Smarter Autonomic Clinic

If you're a clinician looking to upgrade from "I check a standing blood pressure" to "I practice autonomic medicine," here are the essentials:

1. *Beat-to-Beat BP Unit (Finapres, CNAP, Task Force Monitor)*
The difference between "nothing is wrong" and "your baroreflex is asleep."

2. *Tilt Table with Footboard Support*
Because preventing patient collapse should not be an improvisational art.

3. *HRV + Respiratory Coupling Software*
Preferably with artifact correction, not the built-in smartwatch kind.

4. *Sudomotor Analyzer (ESC or QSweat)*
Your early-warning detector for small fiber failure.

5. *Staff Training in Autonomic Protocols*
You cannot run a tilt test the way you run a stress test.

GLOSSARY OF TERMS

(Every term, every abbreviation, every mechanism, every physiologic quirk, every failure mode, every HRV variable, every testing artifact, and every snark-infused definition—all in one place.)

Core Systems and Branches:
ANS—Autonomic Nervous System
The shadow government of human physiology. Runs heart rate, blood pressure, sweating, gut, and temperature while you pretend you're in control.

SNS—Sympathetic Nervous System
The "fight, flight, or why is my heart doing parkour?" branch. Raises heart rate, constricts vessels, boosts blood pressure, and keeps you upright against gravity.

PNS—Parasympathetic Nervous System
The brake pedal. Slows heart rate, promotes digestion, and keeps the system from living in permanent "being chased by a tiger" mode.

CNS—Central Nervous System
Brain + spinal cord. The control tower that sends orders, occasionally misplaces them, and sometimes launches full-body chaos.

Peripheral Autonomic Fibers
The wiring from the spinal cord and ganglia out to organs and vessels. When these fail, the command center is yelling into a broken microphone.

C-Fibers
Tiny, unmyelinated, slow, and fragile autonomic and sensory fibers. First to fail, last to be taken seriously. Control sweating, temperature, microvascular flow, and "burning pins-and-needles" drama.

Small Fiber Neuropathy (SFN)
Injury to small autonomic and sensory fibers. Clinically - weird pain, heat/cold intolerance, patchy sweating, and a neurologist who keeps saying, "Your MRI is normal."

Baroreflex
The reflex loop that adjusts heart rate and vascular tone when blood pressure changes. When it's working, you stand up and do not faint. When it's failing, gravity wins.

Tests, Protocols, and Key Metrics:
HR—Heart Rate
Beats per minute. The main character in many autonomic plots. Goes up, down, or sideways depending on preload, compensation, medication, and emotional drama.

BP—Blood Pressure
The pressure of blood against vessel walls. Measured as systolic/diastolic. Also a favorite place for both physiology and medications to lie to you.

SBP—Systolic Blood Pressure
Top number. Reflects how hard the heart ejects blood during systole. When it collapses on standing, the patient follows.

DBP—Diastolic Blood Pressure
Bottom number. Reflects vascular tone during cardiac relaxation. Helpful for spotting early leaks in the vascular system.

PP—Pulse Pressure
SBP minus DBP. A proxy for stroke volume and preload. Wide PP - plenty of stroke volume. Narrow PP - low volume, low preload, or both. Collapse PP - the floor is calling.

SV—Stroke Volume
The volume of blood the heart ejects with each beat. Low SV = preload failure, compensatory tachycardia, and upright life becoming optional.

HRV—Heart Rate Variability
Beat-to-beat variation in heart rate. The heart–brain text thread. High HRV - adaptive, flexible system. Low HRV - exhausted, overworked, or drug-modified system.

RSA—Respiratory Sinus Arrhythmia
The normal waxing and waning of heart rate with breathing. Practical translation - a built-in stress test for vagal function.

CO_2 / End-Tidal CO_2
Carbon dioxide level at the end of exhalation. Used in autonomic testing to track ventilation, perfusion, and how close someone is to "this feels like I might pass out."

Baseline Testing
Resting measurements of HR, BP, HRV, and other metrics. The starting position before the system is provoked. If the baseline is bad, the stress tests are just insult layered on injury.

Deep Breathing Test
Slow, paced breathing to test parasympathetic function via HRV and RSA. If this is flat, the vagus nerve is phoning it in.

Valsalva Maneuver
Forced exhalation against a closed airway. A four-phase reflex torture test for baroreflex, sympathetic, and parasympathetic function.

Valsalva Phase II
The part where blood pressure should transiently drop and then recover via sympathetic constriction. If it does not recover, sympathetic vasoconstriction is weak.

Valsalva Phase IV/Overshoot
The rebound rise in blood pressure after release. Presence of a clean overshoot = intact reflex arc. Absence = baroreflex failure or major timing dysfunction.

Orthostatic Testing/Tilt Testing
Transition from supine to upright (standing or tilt table) while tracking HR, BP, and symptoms. Tells you how well the system handles gravity without drama.

Orthostatic Intolerance (OI)
Symptom constellation during upright posture - dizziness, brain fog, palpitations, visual tunneling, fatigue, nausea. Tests may show POTS, OH, or normal numbers with absolutely not-normal patients.

Orthostatic Hypotension (OH)
Defined drop in BP on standing. The "I stand up and my vascular system files for bankruptcy" pattern.

Normal–Normal Pattern
Baseline normal, reflexes normal, orthostatic responses normal. Symptoms are likely non-autonomic, intermittent, early, or from another physiologic domain.

Normal–Abnormal Pattern

Baseline is fine, but stress tests (deep breathing, Valsalva, orthostatic) fail. Classic dysautonomia pattern - normal on paper at rest, collapse under real-world conditions.

Abnormal—Abnormal Pattern

Baseline abnormal AND all challenges abnormal. The system is not overreacting; it is failing across the board.

Sudomotor Testing

Tests sweat gland function and small fiber integrity. When sweat fails, it is not being "dramatic"—it is telling you small fibers are dying.

ESC—Electrochemical Skin Conductance

A sudomotor metric reflecting sweat gland ion transport. High = intact fibers; low = neuropathy; asymmetry = localized nerve injury.

QSART—Quantitative Sudomotor Axon Reflex Test

Measures postganglionic sudomotor fibers via stimulated sweating. A nerve mirror for how well small autonomic fibers still answer the phone.

TST—Thermoregulatory Sweat Test

Whole-body sweat mapping under controlled heat. Highlights where sweat is missing, patchy, or theatrically overperforming.

Sudomotor Pattern: Length-Dependent Neuropathy

Feet worse than hands. Classic distal small fiber neuropathy.

Sudomotor Pattern: Global Loss

Hands and feet impaired. Systemic small fiber failure—think metabolic, autoimmune, or neurodegenerative causes.

Sudomotor Pattern: Patchy/Asymmetric Loss
Irregular regions of reduced sweating. Consider spinal pathology, radiculopathy, or localized nerve injury.

Sudomotor Pattern: Normal Sweat + Severe Symptoms
Central dysautonomia. The wires work; the command center is drunk.

Syndromes, Phenotypes, and Mechanisms:
POTS—Postural Orthostatic Tachycardia Syndrome
Excessive HR rise on standing without a significant BP drop. Variants include volume-poor, hyperadrenergic, and neuropathic flavors. All annoying, some very fixable.

Hyperadrenergic POTS
The adrenaline volcano subtype. High NE, wide PP, elevated HR and often BP on standing. Loud tracings, overachieving sympathetic noise, good response to targeted dampening.

Neuropathic POTS
The wiring-problem subtype. Partial sympathetic denervation, especially in the legs. Vessels do not constrict, blood pools, HR compensates, and the patient feels like gravity tripled.

Preload Failure
Not enough blood returning to the heart. Low SV, narrow PP, tachycardia, and upright life that comes with a time limit.

Sympathetic Underactivation
The "my nerves forgot to vasoconstrict" state. Early BP drops, poor orthostatic tolerance, and an SNS that behaves like it is on permanent lunch break.

Sympathetic Excess/Hyperadrenergic State
Too much norepinephrine firing, too often. Wide PP, tremor, sweating, tachycardia, and a test report that looks like caffeine and panic had a baby.

Baroreflex Timing Dysfunction
The baroreflex responds, but late, poorly coordinated, or inconsistently. Produces wild swings, overshoots, undershoots, and clinician confusion.

Neuropathic Autonomic Impairment
Loss of autonomic fibers causes weak reflexes, poor vasoconstriction, reduced HRV, temperature dysregulation, and broadly unimpressed sudomotor function.

Mixed Physiology
The reality for most patients - elements of preload failure + neuropathy + hyperadrenergic compensation + baroreflex issues all at once.

Autonomic Reserve
The system's extra capacity to handle stressors (heat, meals, standing, exercise). High reserve - can participate in life. Low reserve - one grocery trip = two-day recovery.

Medication Classes, Interference Categories, and Shortcuts:
NE—Norepinephrine
Primary sympathetic neurotransmitter. Too little - sludge. Too much - internal earthquake.

Beta-Blockers
Slow HR and blunt NE effects at β-receptors. Brilliant for hyperadrenergic states, a disaster in pure preload failure if overused.

Alpha-2 Agonists (e.g., Clonidine, Guanfacine)
Turn down central sympathetic output. Great for hyperadrenergic chaos; terrible for already underactive autonomic tone.

Midodrine
Peripheral vasoconstrictor. Improves standing BP and reduces pooling when sympathetic tone is weak. Can hide underlying sympathetic failure during testing.

Fludrocortisone
Mineralocorticoid that expands plasma volume by retaining sodium. Good for preload failure; a mask when you are trying to see natural orthostatic responses.

Droxidopa
Norepinephrine precursor. Supports sympathetic tone in underactive systems but needs careful dosing to avoid hyperadrenergic overshoot.

Pyridostigmine (Mestinon)
Acetylcholinesterase inhibitor that enhances parasympathetic signaling. Polishes RSA and baroreflex control; can make a weak vagus look deceptively competent.

SSRIs/SNRIs—Selective Serotonin/Serotonin–Norepinephrine Reuptake Inhibitors
Psych meds that stabilize HR variability and smooth curves. Can "normalize" tracings while hiding underlying autonomic fragility.

TCAs—Tricyclic Antidepressants
Anticholinergic, NE-modulating agents. Flatten parasympathetic responses and can create false patterns of vagal failure.

Glossary of Terms

SGLT2 Inhibitors
Glucose-lowering agents that increase urinary glucose and fluid loss. Combined with diuretics, they can mimic orthostatic hypotension by shrinking volume.

H1/H2 Blockers
Antihistamines used for mast cell stabilization. Can modulate symptoms in suspected mast cell–linked dysautonomia and hyperadrenergic states.

Maskers
Medications that make abnormal physiology look normal. They prop up tone or volume so tracings seem fine while the underlying system is not.

Exaggerators
Medications that make normal physiology look pathologic - stimulants, over-replaced thyroid, heavy caffeine. They create jittery, hyperadrenergic-looking curves.

Hijackers
Drugs that replace normal autonomic responses altogether. The tracings show medication performance, not the patient's physiology.

Saboteurs
Agents that create new dysfunction where none existed. Mimic neuropathy, OH, or vagal collapse out of thin air.

Shape-Shifters
Polypharmacy combinations with mixed receptor effects that generate inconsistent, test-to-test variability and interpretive nightmare fuel.

Treatment, Rehab, and Practical Terms:

Volume Expansion

Raising intravascular volume using salt, fluids, fludrocortisone, and occasionally desmopressin. First-line for preload failure.

Vascular Tone Support

Interventions that support vasoconstriction - midodrine, droxidopa, compression garments, physical counter-maneuvers.

Adrenergic Modulation

Strategies to calm hyperactive sympathetic tone - low-dose beta-blockers, alpha-2 agonists, mast cell stabilization, and sleep/stress interventions.

Parasympathetic Enhancement

Interventions that improve vagal braking - paced breathing, aerobic conditioning, pyridostigmine, and targeted lifestyle changes.

Neural Support & Repair

Strategies to support nerve health - B vitamins, alpha-lipoic acid, anti-inflammatory lifestyle shifts, metabolic optimization, and autoimmune treatment when indicated.

Compression Garments

External venous support - waist-high stockings, leggings, abdominal binders. Lower limb blood is persuaded not to abandon the brain.

Horizontal Conditioning Training (HCT)

Recumbent or semi-recumbent exercise program (rowing, recumbent bike, swimming) used to build stroke volume and tolerance without provoking full orthostatic collapse.

Physical Counter-Maneuvers
Leg crossing, calf pumps, isometric contractions, and other simple muscular tricks that push venous blood back up when gravity is winning.

Mechanistic Classification
Interpretive framework where you label what is actually failing (preload, sympathetic underactivation/excess, neuropathy, baroreflex timing) instead of slapping on a generic dysautonomia label and walking away.

Prognostic Indicators
Test-derived clues predicting trajectory - strength of RSA, presence of Valsalva overshoot, stability of PP, sudomotor pattern, heat tolerance, postprandial stability, stress sensitivity, and response to interventions.

Abbreviation Cheat Sheet (Quick-Scan):
For the "I just need the codebook" moments:
- ANS—Autonomic Nervous System
- SNS—Sympathetic Nervous System
- PNS—Parasympathetic Nervous System
- CNS—Central Nervous System
- HR—Heart Rate
- BP—Blood Pressure
- SBP—Systolic Blood Pressure
- DBP—Diastolic Blood Pressure
- PP—Pulse Pressure
- SV—Stroke Volume
- HRV—Heart Rate Variability
- RSA—Respiratory Sinus Arrhythmia
- CO_2—Carbon Dioxide (often end-tidal in testing)

- OI—Orthostatic Intolerance
- OH—Orthostatic Hypotension
- POTS—Postural Orthostatic Tachycardia Syndrome
- SFN—Small Fiber Neuropathy
- ESC—Electrochemical Skin Conductance
- QSART—Quantitative Sudomotor Axon Reflex Test
- TST—Thermoregulatory Sweat Test
- NE—Norepinephrine
- HCT—Horizontal Conditioning Training
- TCAs—Tricyclic Antidepressants
- SSRIs—Selective Serotonin Reuptake Inhibitors
- SNRIs—Serotonin–Norepinephrine Reuptake Inhibitors
- SGLT2—Sodium–Glucose Cotransporter-2 (inhibitors)

APPENDIX
THE RAPID-REFERENCE ARSENAL

(Because no one has time to search 350 pages while a test report is staring you in the face.)

This appendix gathers every high-yield table, algorithm, interpretation cue, and mechanism map into a single command center.

Use it as your study guide, clinic cheat sheet, pattern recognition bootcamp, and "please don't misinterpret another Valsalva" safety device.

If the main text is the battlefield, this appendix is the weapons locker.

APPENDIX A—AUTONOMIC TEST REFERENCE TABLES
The tables clinicians wish every autonomic text included but somehow never do.

A1. Normal Test Values & Abnormal Patterns
Heart Rate Variability (HRV)

Metric	Normal Meaning	What Abnormal Means
SDNN	Global variability; reserve	Low → Reduced autonomic flexibility; multisystem compromise
RMSSD	Parasympathetic integrity	Low → Vagal weakness; poor braking
HF Power	Vagal tone	Low → Parasympathetic decline (not "stress")

LF Power	Baroreflex + mixed modulation	Low → Baroreflex weakness; timing issues
LF/HF Ratio	Branch balance	High → Sympathetic dominance *or* overcompensation
Total Power	Overall autonomic energy	Low → Fragile physiology

Deep Breathing Test

Pattern	Interpretation
Smooth oscillation ≥15 bpm	Strong vagal function
Low amplitude	Early parasympathetic decline
No oscillation	Severe vagal failure
Irregular pattern	Central timing dysfunction
Excessive HR rise	Sympathetic contamination

Valsalva Maneuver

Phase / Feature	Normal Pattern	Abnormal Pattern	Meaning
Phase II HR rise	Rapid	Minimal	Sympathetic impairment
Phase II BP recovery	Strong	Absent	Vascular incompetence
Phase IV overshoot	Crisp	Blunted	Baroreflex failure
HR braking	Immediate	Delayed	Vagal weakness

APPENDIX

Stand/Tilt Table Test

Parameter	Normal	Abnormal	Mechanism
HR increase	10–20 bpm	≥30 bpm (POTS)	Compensation for low stroke volume
Systolic BP	Stable	Drop ≥20	Orthostatic hypotension
Pulse pressure	30–40 mmHg	Narrow (<25)	Low stroke volume
Symptoms	None	Presyncope / heat / tremor	Perfusion deficits

Sudomotor/ESC

Result	Meaning
High conductance	Intact small fibers
Low feet only	Early length-dependent neuropathy
Low hands + feet	Systemic small fiber neuropathy
Asymmetric loss	Radiculopathy or focal nerve injury
Normal + severe symptoms	Central dysfunction likely

APPENDIX B—PATTERN RECOGNITION: "IF YOU SEE X, THINK Y"

Because symptoms lie, but physiology never does.

Finding	Likely Mechanism	Diagnosis to Consider
Low RMSSD + normal LF	Parasympathetic impairment	Early vagal failure
Large HR rise + stable BP	Hyperadrenergic compensation	Hyper-POTS
Narrow pulse pressure	Low stroke volume	Hypovolemia, pooling

Heat intolerance + sweat loss	Small fiber dysfunction	SFN, autoimmune
Deep breathing flatline	Vagal collapse	Advanced dysautonomia
Absent Valsalva Phase IV	Baroreflex failure	Central dysfunction
Normal sweat + severe symptoms	Central instability	Brainstem-level dysfunction

APPENDIX C—MEDICATION INTERFERENCE COMPENDIUM

Because nothing ruins an autonomic study faster than pharmacology wearing a fake mustache. This section is your quick-reference antidote to misinterpretation.

C1. Medication Interference Master Table

Medication Class	What It Does to Physiology	How It Distorts Interpretation
Beta-blockers	Slows HR, narrows HR variability	Masks POTS; mimics vagal strength
Calcium channel blockers	Blunts vascular tone	Mimics orthostatic hypotension
SNRIs/SSRIs	Raises sympathetic tone	Inflates LF/HF ratio
Stimulants	Raises HR + LF	Mimics hyperadrenergic POTS
Anticholinergics	Suppresses HF/RMSSD	False vagal impairment
Benzodiazepines	Boosts HF	Cosmetic parasympathetic bump
Pressors (midodrine)	Increases vascular tone	Masks autonomic failure
Fludrocortisone	Expands volume	Artificially stabilizes BP

Ivabradine	Lowers HR	Eliminates HR-based POTS criteria
Diuretics	Reduces volume	Mimics orthostatic intolerance
Nitrates/Vasodilators	Drop BP	False-positive OH

C2. "Do Not Test While On These" List

Unless safety requires continuation.

Medication	Why It Ruins the Study
High-dose beta-blockers	Eliminates HR response entirely
Anticholinergics	Erases vagal signal
Strong SNRIs	Mimics hyperadrenergic states
Midodrine (within 12 hr)	Masks vascular tone failure
Fludrocortisone (recent dose)	Normalizes BP artificially
Ivabradine	Removes HR data as a diagnostic tool

If discontinuation isn't safe, interpret around the medication, not through it.

C3. Mechanism → Signature → Adjustment

Mechanism	Medication Signature	Interpretation Adjustment
Low vagal tone	Flat HF/RMSSD	Confirm via deep breathing
Hyperadrenergic state	Elevated LF/HF	Rule out stimulants/SNRIs
Low vascular tone	Wide pulse pressure	Check for vasodilators
Baroreflex delay	Sluggish Phase II/IV	Verify medication timing

APPENDIX D—FLOWCHARTS & DECISION ALGORITHMS

D1. Autonomic Interpretation Master Algorithm

        START
          ↓
  Evaluate Resting State
          ↓
    HRV? HR? BP?

- Instability = global risk

          ↓
  Deep Breathing Test

- Amplitude? Timing?
- If flat → vagal impairment

          ↓
  Valsalva Maneuver

- Phase II? Phase IV?
- If missing = baroreflex/vascular failure

          ↓
    Stand/Tilt Test

- HR rise? BP trends?
- Compensation vs failure

          ↓
    Integration

- Vagal vs baroreflex vs vascular vs central

          ↓
  Identify Failure Mode

- Hyperadrenergic?
- Neuropathic?
- Central?
- Hypovolemic?

          ↓
      PLAN

APPENDIX

D2. POTS Subtype Identification Flowchart

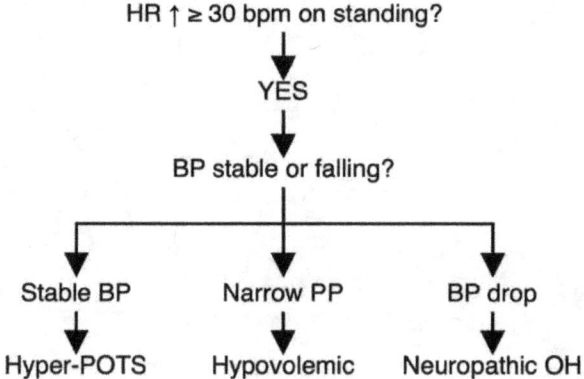

D3. Mechanism-Directed Treatment Selector

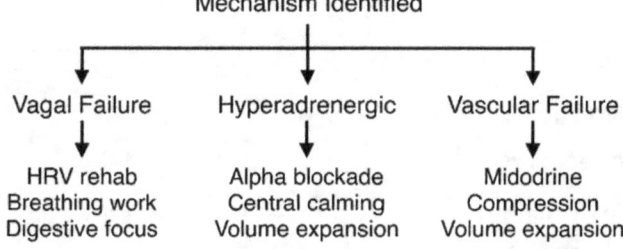

APPENDIX E—GRAPH & CURVE TEMPLATES

Stylized schematics for conceptual comparison only. Not derived from patient data and not intended for diagnostic interpretation. Use real tracings and clinical context for interpretation.

E1. Deep Breathing Curves
Normal Curve:
- Smooth, sinusoidal wave
- ≥15 bpm difference inhalation → exhalation

Vagal Impairment:
- Low amplitude
- Blunted waves

Central Dysfunction:
- Irregular timing
- Asymmetric peaks

E2. Valsalva Curves

Normal 4-Phase Pattern:
- Phase II dip → recovery
- Phase IV crisp overshoot

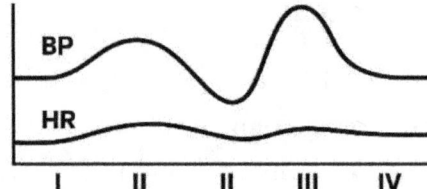

Baroreflex Failure:
- No overshoot
- No HR braking

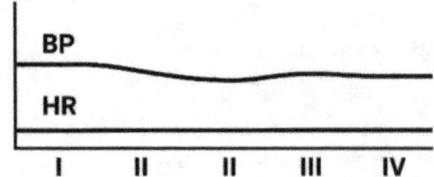

APPENDIX

Sympathetic Under-activation:
- Minimal HR rise
- Poor Phase II recovery

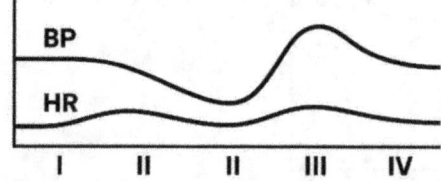

E3. Upright HR/BP Response Templates
POTS:
- HR climbs rapidly
- BP stable
- Pulse pressure narrows

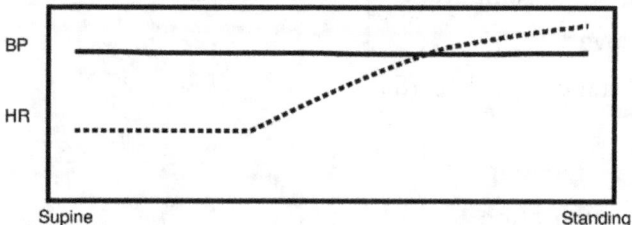

Orthostatic Hypotension:
- BP falls
- HR rises slowly or insufficiently

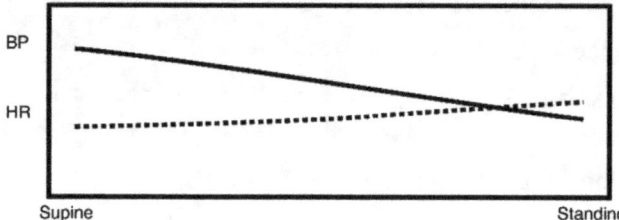

Hyperadrenergic Pattern:
- HR and BP both rise
- Tremor correlates

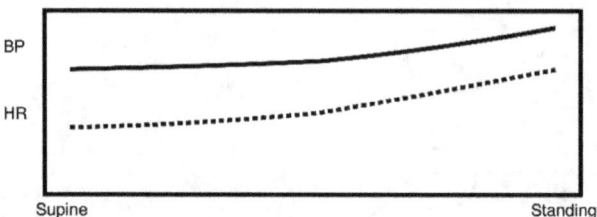

APPENDIX F—MECHANISM MAPS
Because when physiology fails, it fails in recognizable shapes.

F1. Vagal Failure Map
- Low HRV
- Poor deep breathing amplitude
- Slow HR recovery
- Exaggerated standing tachycardia

F2. Baroreflex Dysfunction Map
- Delayed BP stabilization
- Abnormal Valsalva II & IV
- Labile BP at rest

F3. Small Fiber Neuropathy Map
- Declining sudomotor ESC
- Heat intolerance
- Distal sensory changes

F4. Central Dysfunction Map
- Normal sweat
- Abnormal timing patterns
- Unpredictable crashes

APPENDIX G—CLINICIAN CHEAT SHEETS

G1. 15-Second Pattern Sorter

If you see...	Think...	Because...
Huge HR rise, stable BP	Compensation	Low stroke volume
Normal HRV but severe symptoms	Central	Compensation masking
Heat collapse	Small fiber	Poor thermoregulation
BP wobble at rest	Baroreflex	Timing disorder

G2. Red Flags Requiring Specialist Referral

- Absent Valsalva Phase IV
- Sudden-onset severe orthostatic hypotension
- Rapidly declining sudomotor function
- Combined ANS + sensory neuropathy symptoms
- Unexplained supine hypertension

APPENDIX H—PATIENT QUICK REFERENCES

Because understanding your physiology should not require a PhD, a tilt table, or the ability to do math while dizzy.

This appendix is the patient-facing survival kit—the distilled guidance that helps you make sense of your results, your triggers, your symptoms, and your daily patterns without needing to interpret full autonomic tracings or memorize baroreflex physiology.

Use it as your symptom decoder ring, test preparation guide, "Why do I feel like this?" translator, real-world stability toolkit, and mechanism cheat sheet.

If the rest of the book teaches clinicians how to interpret your nervous system, this appendix teaches you how to live with it.

H1. What Each Autonomic Test Measures (In Human Terms)

A simple guide to what your clinicians are actually looking for—and why the tests make you feel the way they do.

1. Deep Breathing Test—"How well does your brake pedal work?"

What you do: Breathe slowly in and out for one minute.

What it measures: Your parasympathetic system (the calming branch that stabilizes your heart and helps you recover).

If results are abnormal, it means:
- Your "rest and digest" system is tired or underperforming.
- Stress, meals, heat, and standing will hit harder.
- Recovery after activity is slower.

In plain language, your body has trouble switching out of emergency mode.

2. Valsalva Maneuver—"Can your wiring handle stress?"

What you do: Blow into a tube or bear down for 15–20 seconds (the world's worst party trick).

What it measures: Your sympathetic system (the accelerator) and your baroreflex (the autopilot stabilizer for blood pressure).

If results are abnormal, it means:
- Your blood pressure overcorrects, undercorrects, or refuses to cooperate.
- You may get dizzy or wiped out from minor stressors.
- Your system may panic too little or too much.

In plain language: Your "fight-or-flight" system and your "stay-upright autopilot" are out of sync.

3. Standing or Tilt Test—"What happens when gravity joins the conversation?"
What you do: Either stand still or get tilted upright like a polite vampire.

What it measures: How your system maintains blood flow to the brain under gravity.

If results are abnormal, it means:
- Your heart is working overtime (POTS).
- Your blood pressure drops too much (orthostatic hypotension).
- Your system gives up (neurocardiogenic syncope).

In plain language, standing up is a physiologic obstacle course, and your nervous system forgot to train for it.

4. Sudomotor Testing—"Do your sweat glands still have Wi-Fi?"
What you do: Place hands/feet on sensors. No poking, no needles, no drama.

What it measures: Small nerve fibers that control sweating and temperature.

If results are abnormal, it means:
- Heat intolerance, shower collapses, and temperature swings have an explanation.
- Early neuropathy is possible.

In plain language, your small fibers are sending "Out of Office" replies.

H2. How To Prepare For Autonomic Testing (And Avoid False Results)

No heroics. No fasting on Everest. Just the essentials.

48–72 Hours Before Testing:
- Avoid alcohol
- Avoid unusually intense exercise
- Maintain hydration
- Avoid experimenting with new medications or supplements

Day of Testing:
- Eat a light meal unless instructed otherwise
- Avoid caffeine unless your clinician says otherwise
- Wear comfortable clothing (gravity is enough of a challenge)
- Bring water, salty snacks, and whatever recovery items you rely on

Medications That Affect Results:
(Some may need to be held; never stop without medical guidance.)
- Beta-blockers
- SNRIs/SSRIs
- Stimulants
- Vasodilators
- Midodrine
- Fludrocortisone
- Anticholinergics
- Ivabradine

What NOT to worry about:
- Breathing "wrong"—you can't
- Failing the test—the point *is* to expose the problem

- Symptoms during testing—clinicians want the physiology to reveal itself

H3. What Your Results *Actually* Mean

This section translates physiology into real-world logic.

If deep breathing is impaired → Your "calming system" is weak.

Expect:
- Quick tachycardia
- Slow recovery
- Morning crashes
- Trouble with meals, heat, stress

If Valsalva is impaired → Your wiring struggles under stress.

Expect:
- Dizziness with exertion
- Blood pressure swings
- Shaking, sweating, or feeling "off"

If standing/tilt testing is abnormal → Your system cannot maintain blood flow upright.

Expect:
- Tachycardia (POTS)
- BP drops (OH)
- Fainting, presyncope
- "Tilt crashes" after meals or showers

If sudomotor testing is reduced → Your small fibers are struggling.

Expect:
- Overheating
- Flushing or blotchy skin
- Intense fatigue after heat exposure
- Burning or tingling sensations

H4. Symptom → Mechanism Quick Decoder

Because "I feel bad" doesn't tell you why.

Symptom	Likely Mechanism	Why
Fast heart rate on standing	Low stroke volume or compensation	Heart trying to keep brain perfused
Heat intolerance	Small fiber dysfunction	Poor sweat/temperature control
Dizziness after meals	Blood flow redirected to gut	Reduced cerebral perfusion
Brain fog	Low perfusion or autonomic instability	Brain is under-fueled
Shaking/tremors	Hyperadrenergic state	Excess norepinephrine
Crushing fatigue after mild activity	Poor autonomic recovery	System never downshifts
Tunnel vision on standing	Low BP or low pulse pressure	Blood not reaching cortex

H5. Triggers And How To Manage Them (Without Losing Your Sanity)

1. Heat
What happens: Blood vessels dilate → blood pools → brain gets less flow.
What helps:
- Cooling vests
- Ice packs
- Avoid hot showers (lukewarm is the new luxury)
- Hydration + electrolytes

2. Meals
What happens: Blood shifts to the gut → perfusion decreases → symptoms spike.

What helps:
- Smaller, more frequent meals
- Lower carb load
- Sit or recline after eating

3. Standing Still
What happens: Muscles stop pumping blood → pooling increases.

What helps:
- Fidget
- Shift weight
- March in place
- Compression gear

4. Exercise

What happens: Demand increases → recovery may lag hours or days.

What helps:
- Recumbent exercise first
- Build tolerance slowly
- Avoid heat exposure during activity

5. Stress

What happens: Autonomic systems misinterpret the signal as "incoming disaster."

What helps:
- Controlled breathing
- Structured pacing
- Reducing sensory overload

H6. The Daily Stability Checklist

A quick way to know whether today is an upright day, a leaning day, or a horizontal day.

Ask yourself:
- How was my morning heart rate?
- Did I wake up exhausted?
- Has heat or humidity changed?
- Am I hydrated?
- Did I sleep poorly?
- Am I fighting an infection?

If three or more are "yes," today is a low-reserve day. Adjust expectations accordingly. Gravity will notice.

H7. When To Call Your Clinician (And Probably 911)
Not every bad day is an emergency. These are.

Call sooner rather than later if you experience:
- Passing out without warning
- New or sudden severe BP drops
- Dramatic worsening heat intolerance
- Burning/numbness that progresses quickly
- Very high heart rate at rest (>120)
- Chest pain (always take seriously)
- Sudden supine hypertension

H8. What To Say At Appointments (So You Get Actual Help)
Instead of saying, "Everything makes me dizzy." → Say, "When I stand, my heart rate increases by around 30 beats within the first minute, and I get tunnel vision."

Instead of "I feel tired all the time." → Try: "My recovery after activity is delayed, especially after heat or meals."

Instead of "I get overwhelmed." → Try: "Stimulus load worsens autonomic stability; noise and heat trigger symptoms."

Clinicians love mechanisms. Give them mechanisms, and you get solutions—not shrugs.

H9. The Patient's Autonomic Toolbox
These are not cures. This is physics.

Core Tools:
- Electrolytes + fluids
- Compression garments

- Cooling devices
- Salt loading (if recommended)
- Activity pacing
- Reclined/recumbent exercise
- Symptom tracking apps or HR monitors
- Recognizing early warning signs

The Most Important Tool—Pattern Recognition.

Your body is not random; it is running a failing algorithm that becomes predictable once you know the inputs.

H10. Final Word: You Are Not "Complicated." You Are Physiologically Outnumbered.

Your symptoms are not personality traits. They are not dramatic. They are not "stress responses." They are the mathematically predictable consequences of a nervous system that is working twice as hard to do half as much.

Understanding the pattern is the first step to reclaiming control. And with the right strategies, the right clinicians, and a bit of well-placed snark, you absolutely can.

www.ingramcontent.com/pod-product-compliance
Lightning Source LLC
LaVergne TN
LVHW021956060526
838201LV00048B/1588